ADVANCES IN
ACCOUNTING EDUCATION
TEACHING AND CURRICULUM INNOVATIONS

Volume 2

ADVANCES IN ACCOUNTING EDUCATION
TEACHING AND CURRICULUM INNOVATIONS

Editors: BILL N. SCHWARTZ
Department of Accounting
Virginia Commonwealth University

EDWARD KETZ
Smeal College of Business Administration
Pennsylvania State University

VOLUME 2

JAI PRESS INC.
Stamford, Connecticut

CONTENTS

LIST OF CONTRIBUTORS

Mohamed E. Bayou

Department of Accounting and
    Finance
University of Michigan, Dearborn

Jean C. Bedard

College of Business Administration
Northeastern University

James L. Bierstaker

Department of Accounting
University of Massachusetts, Boston

Stanley F. Biggs

Accounting Department
University of Connecticut

Robert Bloom

Boler School of Business
John Carroll University

Julia K. Brazelton

School of Business Administration
College of William and Mary

Paul M. Clikeman

E. C. Robins School of Business
University of Richmond

Mary S. Doucet

Department of Finance and
    Accounting
California State University, Bakersfield

Thomas A. Doucet

Department of Finance and
    Accounting
California State University, Bakersfield

Barbara J. Eide

Department of Accountancy
University of Wisconsin, La Crosse

Patricia A. Essex

Department of Accounting and
    Management Information Systems
Bowling Green State University

Gary M. Grudnitski

School of Accountancy
San Diego State University

Daryl M. Guffey   School of Accountancy and
                                   Legal Studies
                                   Clemson University

James M. Kurtenbach   Department of Accounting and
                                   Finance
                                   Iowa State University

Mark W. McCartney   Department of Accounting
                                   Saginaw Valley State University

Alan Reinstein   Department of Accounting
                                   Wayne State University

Robert H. Sanborn   E.C. Robins School of Business
                                   University of Richmond

David Schirm   Boler School of Business
                                   John Carroll University

Bill N. Schwartz   Department of Accounting
                                   Virginia Commonwealth University

James D. Stice   School of Accountancy and
                                   Information Systems
                                   Brigham Young University

Kevin D. Stocks   School of Accountancy and
                                   Information Systems
                                   Brigham Young University

Scott L. Summers   School of Accountancy and
                                   Information Systems
                                   Brigham Young University

John T. Sweeney   School of Accounting Information
                                   Systems and Business Law
                                   Washington State University

W. Darrell Walden   E. C. Robins School of Business
                                   University of Richmond

Carel M. Wolk   Department of Accounting and
                                   Administrative Services
                                   University of Tennessee, Martin

EDITORIAL REVIEW BOARD

EDITORIAL POLICIES

1. You should type and double-space manuscripts on 8 1/2" by 11" white paper. You should use only one side of the page.
2. To be assured of an anonymous review, authors should NOT identify themselves directly or indirectly in the text of the paper. Authors should avoid reference to unpublished working papers and dissertations. If necessary, authors may indicate the reference is being withheld for the reasons cited here.
3. Authors ultimately must submit accepted manuscripts on an IBM compatible disk.
4. Authors should not submit manuscripts currently under review by other publications or manuscripts that have already been published (including proceedings from regional or national meetings). Please include a statement to that effect in the cover letter accompanying your submission. Authors can submit complete reports of research presented at a regional or national meeting that they did not publish in the proceedings.
5. Authors should submit **THREE** copies of each manuscript. Empirical manuscripts go to J. Edward Ketz at The Pennsylvania State University. Authors should include copies of all research instruments. Non-empirical manuscripts go to Bill N. Schwartz at Virginia Commonwealth University.
6. The author should send a check for $35 made payable to **Accounting Education** each submission, whether it is the initial submission or a revision.

STATEMENT OF PURPOSE

Advances in Accounting Education is a refereed, academic research annual whose purpose is to meet the needs of individuals interested in the educational process. We plan to publish thoughtful, well-developed articles that are readable, relevant and reliable.

Articles may be non-empirical or empirical. Our emphasis is pedagogy, and articles **must** explain how teaching methods or curricula/programs can be improved.

Non-empirical manuscripts should be academically rigorous. They can be theoretical syntheses, conceptual models, position papers, discussions of methodology, comprehensive literature reviews grounded in theory, or historical discussions with implications for current and future efforts. Reasonable assumptions and logical development are essential. Most manuscripts should discuss implications for research.

For empirical reports sound research design and execution are critical. Articles should have well articulated and strong theoretical foundations. In this regard, establishing a link to the non-accounting literature is desirable. Replications and extensions of previously published works are encouraged. As a means for establishing an open dialogue, responses to or comments on articles published previously are welcomed.

REVIEW PROCEDURES

Advances in Accounting Education will provide authors with timely reviews clearly indicating the review status of the manuscript. The results of initial reviews normally will be reported to authors within **eight** weeks. Authors will be expected to work with an Editor who will act as a liaison between the authors and the reviewers to resolve areas of concern.

WRITING GUIDELINES

1. Each paper should include a cover sheet with names, addresses, telephone numbers, fax numbers, and e-mail address for all authors. The title page also should include an abbreviated title you should use as a running head (see item #six below). The running head should be no more than 70 characters, which includes all letters, punctuation and spaces between words. You should not identify yourself anywhere else in the manuscript.

2. Manuscripts should include on a separate lead page an abstract not exceeding 250 words. The author's name and affiliation should NOT appear on the abstract. It should contain a concise statement of the purpose of the manuscript, the primary methods or approaches used (if applicable), and the main results, conclusions, or recommendations.

3. You should begin the first page of the manuscript with the manuscript's title. DO NOT use the term "Introduction" or any other term at the beginning of the manuscript. Simply begin your discussion.

4. Use uniform margins of 1 1/2 inches at the top, bottom, right and left of every page. Do not justify lines, leave the right margins uneven. *Do not* hyphenate words at the end of a line; let a line run short or long rather than break a word. Type no more than 25 lines of text per page.

5. Double space *among all* lines of text, which includes title, headings, quotations, figure captions, and all parts of tables.

6. After you have arranged the manuscript pages in correct order, number them consecutively, beginning with the title page. *Number all pages.* Place

the number in the upper right-hand corner using Arabic numerals. Identify each manuscript page by typing an abbreviated title (header) above the page number.

7. We prefer **active** voice. Therefore, you can use the pronouns "we" and "I." Also, please avoid using a series of prepositional phrases. We strongly encourage you to use a grammar and spell checker on manuscripts before you submit to our journal.

8. All citations within your text should include page numbers. An appropriate citation is Schwartz (1994, 152) or Ketz (1995, 113-115). You do not need to cite six or seven references at once, particularly when the most recent references refer to earlier works. Please try to limit yourself to two or three citations at a time, preferably the most recent ones.

9. You should place page numbers for quotations along with the date of the material being cited. For example: According to Beaver (1987, 4), "Our knowledge of education research . . . and its potential limitations for accounting. . . ."

10. *Headings*: Use headings and subheadings liberally to break up your text and ease the reader's ability to follow your arguments and train of thought. First-level headings should be upper-case italics, bold face, and flush to the left margin. Second level headings should be in bold face italics, flush to the left margin with only the first letter of each primary word capitalized. Third-level headings should be flush to the left margin, in italics (but not bold face), with only the first letter of each primary word capitalized.

11. You should list any acknowledgments on a separate page immediately after your last page of text (before the *Notes* and *References* Sections). Type the word "Acknowledgment," centered, at the top of a new page; type the acknowledgment itself as a double-spaced, single paragraph.

12. You should try to incorporate endnote/footnote material into the body of the manuscript. When you have notes, place them on a separate section before your references. Begin notes on a separate page, with the word "Notes" centered at the top of the page. All notes should be double-spaced; indent the first line of each note five spaces.

13. Your reference pages should appear immediately after your "Notes" section (if any) and should include only works cited in the manuscript. The first page of this section should begin with the word "References" centered on the page. References to working papers are normally not appropriate. All references must be available to the reader; however, reference to unpublished dissertations is acceptable.

14. You should label TABLES and FIGURES as such and number them consecutively (using Arabic numerals) in the order in which you mention them first in the text. Indicate the approximate placement of each table/figure by a clear break in the text, inserting:

TABLE (or FIGURE) 1 ABOUT HERE

Set off double-spaced above and below. Tables should be placed after your References section: figures should follow tables. Double-space each table/figure and begin each on a separate page.

15. Parsimony is a highly desirable trait for <u>manuscripts</u> we publish. Be <u>concise</u> in making your points and arguments.

16. <u>Sample Book References</u>

Runkel, P. J. and J.E. McGrath. 1972. *Research on human behavior. A Systematic guide to method.* New York: Holt, Rinehart and Winston.

Smith, P. L. 1982. Measures of variance accounted for: Theory and Practice. Pp. 101-129 in *Statistical and methodological issues in psychology and social science research,* edited by G. Keren. Hillsdale, NJ: Erlbaum.

17. <u>Sample Journal References</u>

Abdolmohammadi, M. J., K. Menon, T. W. Oliver, and S. Umpathy. 1985. The role of the doctoral dissertation in accounting research careers. *Issues in Accounting Education*: 59-76.

Thompson, B. 1993. The use of statistical significance tests in research: Bootstrap and other methods. *Journal of Experimental Education* 61: 361-377.

Simon, H.A. 1980. The behavioral and social sciences. *Sciences* (July): 72-78.

Stout, D. E. and D. E. Wygal. 1994. An empirical evidence of test item sequencing effects in the managerial accounting classroom: Further evidence and extensions. Pp. 105-122 in *Advances in Accounting*, vol. 12, edited by Bill <u>N.</u> Schwartz. Greenwich, CT: JAI Press.

THE DUAL ROLE OF CRITICAL THINKING IN ACCOUNTING EDUCATION

Mohamed E. Bayou and Alan Reinstein

ABSTRACT

Critical thinking has played two roles in accounting education. As a *cause*, critics, by applying critical thinking, have demonstrated many weaknesses in accounting education over the last 100 years and recommended significant changes to be undertaken. The weaknesses include focusing on knowledge acquisition rather than acquisition *and* utilization, over-emphasis on technical procedures, de-emphasis of theory and liberal arts, and promoting the passing of the uniform CPA exam as the prime goal of accounting education. These criticisms have motivated several educational institutions to substantially revise their accounting curricula. When these changes are implemented, critical thinking becomes an *effect*, where it gains a wide acceptance by most accounting educators as a basic skill to be incorporated into the new curriculum.

However, as an *effect*, critical thinking skills lack a general definition in accounting and consist of complex and conflicting concepts that can be difficult to implement. This paper critically examines the concepts of "thinking," "criticality," and "skills" as developed in the fields of philosophy and psychology. From this review, the paper

Advances in Accounting Education, Volume 2, pages 1-20.

extracts the basic features of these concepts consistent with the nature of accounting education. The paper applies the three components to skills several organizations and studies in accounting education have recommended.

THE DUAL ROLE OF CRITICAL THINKING IN ACCOUNTING EDUCATION

Overwhelming evidence accumulated over the last decade suggests that accounting graduates lack critical thinking skills (Federation of Schools of Accountancy (FSA) 1997; Nelson 1995; Siegel and Kulesza 1995; Williams 1993; AAA Bedford Committee, 1986). Nelson (1995, 70) explains that because the CPA exam fails to test "critical thinking, analysis, synthesis and professional judgment, motivation has existed for accounting educators to increase emphasis upon the memorization of accounting rules, rather than the theoretical concepts upon which the rules were based." Several organizations have recommended or required incorporating critical thinking skills into accounting curricula—including the AAA's Bedford Committee; American Institute of Certified Public Accountants (AICPA), Accounting Education Change Commission (AECC), the Institute of Management Accountants (IMA), American Assembly of Collegiate Schools of Business (AACSB) and the then Big Eight CPA firms' white paper (Perspectives 1989). In response, certain universities listed in Table 1 have embarked on curriculum change to cover critical thinking at the undergraduate and graduate levels.

Table 1. Educational Institutions That Have Revised Their Curricula to Incorporate Critical Thinking

At the Undergraduate Level[*]	At the Graduate Level[**]
Arizona State University	Babson College in Massachusetts
Brigham Young University	University of Chicago
University of Illinois	University of Denver
Kansas State University	Harvard Business School
Kirkwood Community College	Indiana University
University of Massachusetts at Amherst	University of Pennsylvania
Mesa Community College	
Nassau Community College	
North Carolina A&T State University	
University of North Texas	
University of Notre Dame	
Rutgers University-Newark	
University of Virginia	

Sources: [*]Williams (1993), and Cassagio (1997).
            [**]Svetcov (1995), and *U.S. News & World Report* (1993).

These developments suggest that critical thinking has a dual role to play in accounting education—as a cause and an effect. As a *cause*, it has driven accounting education to make significant changes and, as an *effect*, critical thinking has become a basic skill in the revised curricula at certain universities. For the effect role, lacking a generally accepted definition in accounting (FSA 1997, 1), critical thinking may mean different things to different educators. Thus, an effective design of these skills remains elusive without properly understanding the nature of critical thinking.

We examine the nature of such thinking and show how various skills can promote critical thinking. Our first section explains critical thinking's dual role in accounting education. We next explore and apply the concepts of "thinking," "criticality," and "skills" to help explain how the new skills can apply critical thinking in accounting which is the third section. Problems, pitfalls, and insights into applying critical thinking skills follow these applications in the fourth section. Finally, we present a summary and conclusions. All of the examples, except the referenced ones, are based on the authors' experiences.

CAUSE AND EFFECT OF CRITICAL THINKING IN ACCOUNTING EDUCATION

Critical thinking's dual role in accounting education takes the form of the (1) *cause* that has created the urgent need for change; and, (2) *effect* (product) of the implemented change. We explain these dynamics as follows.

Relationship between Accounting Knowledge and Critical Thinking

Over the past few decades, such authors as Previts and Merino (1979), Langenderfer (1987), Zeff (1989), Williams (1993), Nelson (1995), and FSA (1997) sharply criticized the development of accounting knowledge. They recommended significant changes in accounting knowledge at two stages: knowledge *acquisition* and knowledge *utilization*.

Accounting Knowledge Acquisition

Knowledge acquisition pertains to the input side of accounting education characterized by qualitative and quantitative dimensions of this knowledge. Historians argue that the first 65 years of the twentieth century emphasized teaching *technical knowledge* coupled with the primary goal of passing the uniform CPA examination (Williams 1993; Nelson 1995). According to Previts and Merino (1979, 154), after securing acceptance of accounting curricula in universities at the beginning of this century, accountants began promoting broader goals and more conceptual programs. Practitioners believed education's role was to develop analytical ability and emphasize theory and philosophy, while mastering technical accounting and auditing procedures are better learned in the field after graduation (Previts and Merino 1979, 154). Their disappointment grew as educators increasingly emphasized procedural orientation and a kind of vocationalism at the expense of reducing liberal education (Previts and Merino 1979, 154). According to Langenderfer (1987, 304), accounting education has not fulfilled the expectations of the profession's current or past leaders. From 1920 to 1960, emphasis on preparing students for the CPA examination became one of the "hallmarks" of accounting education (Langenderfer 1987, 308). During this period, schools competed against each other on the "pass rate" basis of their graduates (AAA 1986, 172).

Disappointment with this state of accounting education is manifested in several criticisms. For example, several prestigious groups (AAA 1967; Perspectives 1989; AECC 1991) criticized this state and urged separating the CPA examination from the accounting curriculum (Nelson 1995, 64). Zeff (1989, 203) laments that accounting has been reduced to "a collection of rules that are to be memorized in an uncritical, almost unthinkable way," and he questions whether accounting as it is presently taught properly occupies a place in the university. Cunningham (1996, 50) warns that "More creative students and those who are comfortable in ambiguous situations tend to choose other majors where they believe they can better use those characteristics."

While the qualitative dimension of accounting knowledge primarily emphasized technical procedures and passing the CPA exam, three major factors drove its *quantitative* dimension: (1) the sheer quantity of authoritative pronouncements (e.g., those of the Accounting Principles Board, Financial Accounting Standards Board, AICPA, and Securities and Exchange Commission); (2) the fixed number of credit hours available for teaching (Nelson 1995, 64); and (3) educators' belief that they should teach every technical aspect of accounting procedures (Nelson 1995, 64; Williams 1993, 80). As the common body of knowledge expanded, educators responded by adding more specialized accounting courses (Previts and Merino 1979, 154) that further reduced the liberal arts component (Nelson 1995, 64). As the breadth of education further narrowed to sacrifice depth, courses turned into "a funnel to pour information into a student" (Nelson, 1995, 1989). Keating and Jablonsky (1990, 62) neatly summarize this state of affairs as follows:

> The proliferation of statements of financial accounting standards has had the insidious effect of pushing the accounting curriculum in an ever more technical and ever less business-oriented direction. The curriculum has become devoted to teaching students the technical rules and conventions of conformance. More and more it concentrates on formal accounting rules, with correspondingly less focus on essential business and social issues.

Thus, accounting education created the need for change from a technical orientation, vocationalism, and CPA-exam focus to embracing broader goals and inte-

Table 2.   New Approaches, Objectives, and Methods of
Accounting Education

New Approaches[1]	New Objectives[1]	New Methods[2]
• Broader emphasis on general education business, and organizational knowledge	• Present views in writing • Present views through oral presentations	• Internet-based research • Computer simulations
• Heavy integration of accounting courses	• Read, critique, and judge the value and contribution of work presented	• Electronic spreadsheet projections
• Increasing emphasis on unstructured problems		• Class presentation
• Increasing emphasis on the learning process	• Listen effectively	• Group projects
• Adding broader objectives	• Work in and with teams	• Professional writing
• Increasing emphasis on communication skills	• Solve diverse and unstructured problems	• Exposure to interdiscipline team in unfamiliar settings teaching
• Increasing emphasis on active student participation	• Deal with imposed pressures	• Journal article outlining
• Use and integration of technology in accounting curriculum	• Organize and delegate tasks	• Case analysis
• Accounting principle courses focus on:	• Resolve conflicts	• Focused critiques
• Accounting's role in society		• Field trips
• Accounting's role in organizations		• Video presentations
• Accounting's role in decision making		

Sources: [1] Williams (1993, 82)
[2] Various sources including Armenic (1985, 68), Williams (1993, 76-81), Cassagio (1997, 13), and the author's own experiences.

gration of knowledge. Several recommendations for adopting new approaches, new objectives and new methods (see Table 2) capture the essence of this change.

Table 2 presents new approaches, objectives, and methods to accounting education given this new emphasis on critical thinking skills. For example, such new methods as Internet-based research skills can help fulfill the new objective of solving diverse and unstructured problems in unfamiliar settings, which, in turn, provide new approaches of increasing emphasis on the learning process and greater use of technological skills.

Accounting Knowledge Utilization

Most accounting educators and practitioners now take for granted the need to incorporate critical thinking skills in accounting education curricula (Williams 1993, 76; Deppe et al. 1991, 264). This development marks a major shift from exclusive emphasis on knowledge acquisition to both knowledge acquisition and utilization. Knowledge *utilization* focuses on relating the object of learning to practice. Such widely discussed topics as "teaching accounting in action" rather than in a vacuum (Amernic 1985, 64), "relevance lost" (Johnson and Kaplan 1987), "relevance regained" (Johnson 1992, 55) and "defining the customers of accounting education" (Reckers 1995) exemplify the importance of orienting accounting education toward practice and real-world problems, that is, toward learning how to *utilize* accounting knowledge to help the "customers" of accounting education. However, critical thinking is a complex concept even in the field of philosophy, a matter that Finocchiaro (1990, 1989) and Siegel (1990, 1988) debated for years.

The FSA's report (1997, 1) calls critical thinking "a term with no universally accepted definition." The many, diverse definitions of critical thinking lead Hofer and Pintrich (1997, 89) to find little agreement on the dimensions characterizing individual thinking and reasoning and possible linkages between various relevant theories. The FSA (1997, 4) argues that the diverse views of the nature of thinking "highlight the need for additional research to identify connections between various facets of skills that may be related to critical thinking." The FSA (1997, 6) concludes that "given the lack of agreement on evaluating, studying, or even defining critical thinking, it is premature to make changes in [accounting] education at this point."

In short, accounting has failed to examine thoroughly the concepts of "thinking," "critical," and "critical thinking." As explained below, these concepts can be clarified so that changes in accounting education will be grounded on the recommended changes of the AECC and the other studies cited above.

THE NATURE OF CRITICAL THINKING SKILLS

The phrase "critical thinking skills" consists of complex concepts, which we examine closely by first discussing "thinking," then "critical" and lastly "skills."

Thinking. Idealists and empiricists in philosophy, psychology and other disciplines argue whether thinking is a property of the mind or the body—usually found on opposite extremes of a continuum. Such idealists as the French philosopher René Descartes (commonly called the father of modern philosophy) and the German philosophers Leibniz, Kant and Hegel hold that the mind is a metaphysical entity that interacts with the material body so that thinking is a property of the mind, not the body (Dellarosa 1991, 1). Ericsson and Oliver (1991, 392) explain that thinking "encompasses a huge variety of mental activities. We are thinking when we solve problems, day-dreaming, remember facts, and so on." Johnson-Laird (1991, 429) explains that the dazzling variety of thinking has caused some cognitive scientists to despair of our ever understanding it completely. This introspective view of thinking led many philosophers to believe that

(1) thought is always sequential, that is, the thinking process proceeds into a sequence of images and thoughts, and (2) all humans share the basic characteristics of the mind (Johnson-Laird 1991, 430).

In contrast, such empiricists as Hobbes view thought as a wholly materialistic process. In this mechanistic view, thinking is like computing sums where instead of trafficking in numbers, it requires trafficking in *ideas* (Dellarosa 1991, 1). Accordingly, thinking meant combining ideas to form new ideas, subtracting ideas from each other, comparing ideas, and so on. This view fueled scientific investigation into the nature of thought because, to Hobbes, matter could think, which meant that machines capable of thinking could be built (Dellarosa 1991, 1). Thus, the fact that many technical tasks in accounting can be done entirely on computers is consistent with this "scientific" view of thinking. The IMA's "Project Millennium" (1997, 2) and Keegan and Portik (1995, 26) explain that since new electronic information systems make these tasks simple and economical, the problem is not data collection and information recording; rather, it is how to correlate and integrate the myriad volumes of information in order to discover subtle patterns in the information for top management. This means that the required thinking in accounting has been raised to a higher order—from procedural to critical.

Between these two extreme schools of thought (the idealist or introspective and empiricist) on thinking, Dellarosa (1991), Gardner (1985) and Haugeland (1985) provide historical reviews of several systems and theories on the nature of thinking. For example, the physiological systems, structuralism and behaviorism, which subscribe to the primacy of materialistic view of thought, deny the legitimacy of mental concepts such as thinking (Dellarosa 1991, 6; Ericsson and Oliver 1991, 395). In contrast, the Gestalt psychology opposes this materialistic view by providing several instances where the internal dynamic of thinking significantly influences behavior. The current view on thinking emerges from the cognitive revolution which started in the mid-1960s (Dellarosa 1991, 11). Labeled "cognitive science," this new science embraces research from diverse fields including philosophy, linguistics, psycholinguistics, computer science and neuroscience, and shares the common goal of seeking "explanation of higher mental processes" (Dellarosa 1991, 11).

Given these opposing systems on the nature of thinking, an elaborate design of critical thinking skills becomes problematic. Accepting all of these systems as a foundation for the design is bound to face contradictions and conflicting results. Yet, limiting the design to only a few of them is impractical since (1) different accountants may hold different views about the nature of thinking; (2) integrating accounting with other disciplines requires openness to diverse views; and (3) the "criticality" in critical thinking emphasizes this openness as discussed below.

What is "critical" in critical thinking?   A debate in philosophy developed between Siegel and Finocchiaro regarding the nature of critical thinking. Finocchiaro (1989, 483) objected to Siegel's (1988, 30) equation, critical thinking =

good reasoning = rationality, in that "good" reasoning and rationality need not be critical, that is, they need not involve negative criticism (Siegel 1990, 453). Finocchiaro (1990, 462) argues that Siegel's equivocation ultimately is "reduced to questionable appeal to authority and to question begging." Instead, Finocchiaro (462) defines critical thinking as "the special case of reasoning when explicit reason assessment is present." Siegel (1990, 458) agrees with Finocchiaro in that "critical thinking is not coextensive with rationality." From this debate, it appears that there are three senses of "critical," negative, positive, and neutral (Siegel 1990, 458; Finocchiaro 1990, 462).

One source of their disagreement is their different views of "thinking." Finocchiaro (1990, 463) argues that "reasoning is a special type of thinking, so that while all reasoning is thinking, not all thinking is reasoning," but later qualifies this statement by arguing that since "reasoning is a form of evaluation [...] all thinking is ultimately reasoning." It suffices for this paper's purpose to mention that the debate may settle on the view that "**critical thinking is thinking which is reasoned, evaluative, and self-reflective**" (Finocchiaro 1990, 465, bold added). Johnson-Laird (1991, 454) explains self-reflection as a meta-cognition of a higher-order type of thinking that depends on having access to a model of a thought process that gives rise to self-awareness, as explained in the following section. As to the question "must thinking be critical to be critical thinking?" Finocchiaro (1990, 465) replies:

> I believe that it is probably true that all thinking which is reasoned *and* evaluative *and* self-reflective is critical thinking. Then insofar as reasoned, evaluative, and self-reflective are three senses of "critical," we may also say that critical thinking is, indeed, thinking which is critical.

Effective accounting knowledge utilization, that is, knowledge designed as relevant to target customers, requires developing accounting graduates' reasoning, evaluative, and self-reflective *skills*. For accounting education, Rottenberg (1985), Bayou and Panitz (1993), and May and May (1996) develop reasoning and evaluating skills for *argumentation*, *persuasion*, and *communication*, respectively. Self-reflection is enhanced by switching from passive *teaching* to emphasizing *learning* by involving students in dynamic knowledge acquisition, i.e., learning how to learn (Williams 1993, 76).

Skills. Merriam-Webster's *English Dictionary* (1995, 485) defines *skill* as "the ability to use one's knowledge effectively in doing something." To grasp the nature of this ability, we should distinguish between the function of thinking (*what* the mind does) and the underlying procedures it follows (*how* it does it), a structure that Marr (1982, 27) and Johnson-Laird (1991, 429) developed. The former component (*what* the mind is doing) refers to the "computational level" and the latter (*how* it is doing it) pertains to the "algorithmic level" of thinking

(Johnson-Laird 1991, 429). As discussed below, the computational level is more relevant to critical thinking skills in accounting education.

Most thinking lies between the extremes of *daydreaming* (thinking-wandering with no goal nor purpose) and *calculating*, which has a goal and a global structure (Johnson-Laird 1991, 432). This structure is based on characteristics of the "problem space" Newell and Simon (1972, 19) developed. They pioneered the computational analysis of problem solving and provided a unifying framework for planning and thinking (Johnson-Laird, 1991 432). We briefly analyze this structure below.

In all cases, to reach the goal of solving the problem, there is a *starting point* (initial conditions) and a set of mental *operations* that must be carried out appropriately. There is a "space" of all possible sequences of operations; what has to be worked out is a sequence that forms a route through the space from the initial state to the goal:

$$\text{State } 1 \rightarrow \text{State } 2 \rightarrow \text{Goal}$$

A successful plan generates a route to solve the problem (Johnson-Laird 1991, 432).

According to this analysis, the skill of applying the accounting cycle is "deterministic" because at each point, the next step in the cycle is determined wholly by its current state. Only one route exists through the problem space from one state to the next, and the accountants' knowledge enables them to follow it easily (Johnson-Laird 1991, 432). That is, in accounting education, thinking has a global structure based on characteristics of the problem space which, when learned, develops a "computational skill" defined as "the ability to execute an internal hierarchical organization" (p. 430). Miller et al. (1960, quoted in Johnson-Laird, 1991, 431) and other psychologists call this organization a "plan" which they define as "any hierarchical process in the organization that can control the order in which a sequence of operations is to be performed." They demonstrate convincingly that planning is a major part of thought, a demonstration that "sealed the demise of behaviorism" (432). Developing a skill for the computational-level of thinking in accounting education should focus on helping students recognize the hierarchical process needed in a thinking situation and learn how to plan and perform an efficient sequence of operations that leads to the goal of the thinking endeavor. For critical thinking, there are several types of these computational-level-thinking skills, as presented below.

CRITICAL THINKING SKILLS IN ACCOUNTING EDUCATION

Accounting educators usually focus on the computational level and function of thinking (i.e., *what* the mind is doing [Marr 1982, 22]), since this function is the basis for developing knowledge acquisition and utilization. Upon learning how to execute internal hierarchical organizations, students develop computational-level thinking skills as explained above. In contrast, the algorithmic level (i.e., *how* the mind is thinking) has concerned many psychologists (Johnson-Laird 1991, 429). While psychologists may find the algorithmic level of thinking more interesting, accounting educators may consider the computational level more relevant to develop critical thinking skills. The accounting education literature (Williams 1993, 81; Cassagio 1997, 13) recommends several skills, including the following:

- Problem finding
- Problem solving
- Looking for assumptions
- Hypothesizing
- Hypothesis testing
- Interpreting results
- Summarizing
- Concluding
- Sensitivity (what-if) analysis
- Distinguishing

- Designing projects
- Responding to criticism
- Writing
- Listening
- Oral presentations
- Synthesizing
- Observing
- Comparing
- Classifying

This list includes major skills accounting educators consider essential for accounting education. We demonstrate with examples how some of these skills can promote critical thinking by focusing on Finocchiaro's (1990, 465) three components:

Reasoning: ascertaining "why?" in searching for the causality underlying a claim.

Evaluating: weighing the pros and cons of the causes of the claim.

Self-reflecting: students' awareness of the *incremental* knowledge they learn in a given thinking exercise.

1. *Problem-finding skill.*   This is probably the most important skill for critical thinking (FSA 1997, 12) and for creative thinking (Gronhaug and Kaufmann 1988, 2). According to Gronhaug and Kaufmann (1988, 2), many "creativity" researchers call problem-finding the "hidden nine-tenths of the creativity iceberg." Kabanoff and Rossiter (1994, 288) note that the volume of empirical research devoted to problem-finding is remarkably small and conclude that a better understanding of problem-finding clearly remains as "one of the most vital and difficult frontiers for creativity researchers. It is a 'messy' concept that is hard to define, hard to operationalize, and almost always requires eclectic research methods. Nevertheless, for many theorists, problem-finding is a crucial element of creativity, especially for real-world creativity in applied settings."

Several authors recommend unstructured formats for case analysis that will enable students to "find" and "identify" the core problems in assigned cases

(Williams 1993, 80; Deppe et al. 1991, 283). The *reasoning* dimension of critical thinking in mastering this skill begins with examining apparent symptoms as clues to a structure, for example, inappropriate managerial behavior as a clue to a fraudulent scheme in an organization, excessive scrap as an indication of problems in a manufacturing process, and increasing customer complaints as a sign for quality issues regarding products or services. The *evaluating* component of critical thinking directs students to (1) list the various causes for the problem(s); (2) classify these causes into primary and secondary groups, and, (3) elaborate on the basis for such classification. Finally, *self-reflecting* evolves from students' direct experience and awareness of the difficulties involved in problem finding. To assure that students cover this critical thinking step, an accounting educator may have students summarize the major objectives of the exercise and what they learned from the experience. For example, in a group project on financial-statement analysis, an instructor may ask students to summarize how the project helped them (1) "discover" major problems facing the reporting entity; (2) "read between the lines" to discern unusual relationships in the financial reports; and, (3) decide on the key tools needed for diagnosing certain problems.

2. *Problem-solving skill.* Among the listed skills, problem-solving is probably the most commonly developed in accounting education because most homework assignments, class exams, quizzes, and professional examinations require mastering this skill in order to succeed. Learning this skill requires experiencing the three phases of decision making: screening, preference, and justification. For example, in a capital budgeting replacement of an old equipment decision, students would examine several replacement alternatives, delete the ineligible ones, and list the acceptable choices. This exercise, which requires developing criteria for the screening phase, constitutes the *reasoning* step of critical thinking. The *evaluating* step is the preference decision. That is, to choose the best equipment to replace the old one, students may evaluate the screened alternatives by using several methods, e.g., the present value, payback period, internal rate of return, or the accounting rate of return method. The final decision may need justification by employing selected criteria as these methods or other factors such as urgency, capacity, and regulatory constraints. Finally, the *self-reflecting* step may require students to consider the totality of the problem-solving project to help ascertain the incremental knowledge gained from this experience. For example, the instructor could ask students to write up a plan summarizing the basic sequence of thought processes in certain capital replacement decisions.

3. *Looking-for-assumptions skill.* Probably the first step in any critical thinking endeavor is to address critically the question(s) asked before attempting to answer such question(s). A question may be emotional-laden or loaded with several explicit and hidden assumptions, which must be recognized first in order to place the inquiry in its proper perspective. A major purpose of training students on this skill is to develop a trait of professional skepticism. In auditing, Statement of Auditing Standards (SAS) No. 82 (AICPA, par. 27) defines the term *professional*

Table 3. Demonstrating How New Skills Can Apply the Three Components of Critical Thinking

Skills	Reasoning	Evaluating	Self-reflecting
• Problem finding	Search for what causes complaints, losses, failure, or suffering to occur.	Rank the causes into primary and secondary factors	How can this exercise help you improve your ability to identify similar problems?
• Problem solving	Screen the alternative viable means to deal with the problem.	Using a performance criterion, rank these alternatives into optimum and sub-optimum solutions.	What are the essential procedures necessary to solve this type of problems?
• Looking for assumptions	For every given statement, recognize any explicit and hidden assumptions.	Which assumptions are valid? Invalid?	How can you recognize assumptions and Determine their validity?
• Hypothesizing	What is the purpose of a hypothesis? How should it be formed?	Is a given hypothesis testable or merely an untestable value judgment?	Do you know the characteristics of a statistically testable hypothesis?
• Hypothesis testing	Why is the given method appropriate to test a given hypothesis?	Is this method the best one for testing this type of hypotheses?	What alternative methods can you apply to test this type of hypotheses?
• Interpreting results	Do results need interpretation? For whom?	Does the interpretation cover the basic tenets in an understandable fashion?	Interpreting results depends on the intended receivers of the results.
• Summarizing	What is the purpose of summarizing?	Is the summary too long? Does it capture the basic points?	What should be included in a summary? How Long should be a summary?
• Concluding	Why is a conclusion important?	How can you assess the leap from given results to stated conclusions?	Conclusions differ from summaries; the former adds new information; the latter do not.
• Sensitivity (what-if) analysis	Why is sensitivity analysis important for budgeting?	(1) How many iterations are needed? (2) What criteria do you need to select the best alternative?	Learn by practicing, i.e., use electronic spread-sheets to perform what-if analysis.
• Distinguishing	Distinguish among the following: (1) Statements of fact, value judgment, and policy statements. (2) Facts from assumptions. (3) Well-supported from non-supported claims.	Should a given topic be presented by an accountant in terms of facts, value judgment or a policy statement?	Do you understand the basic differences between these three modes of arguments?

• Designing projects	What are the essential factors for designing an assigned project?	How can one measure the effectiveness of a given project?	How can a given exercise help you design similar projects?
• Responding to criticism	(1) How can you distinguish objective from emotive criticism? (2) Should you respond to each kind? How?	What is the most effective way of responding to a given criticism?	Did an emotive criticism succeed in drawing you into an emotive battle?
• Writing	How does writing in accounting differ from writing in liberal arts?	Assess the quality of writing of a given accounting article.	How does the given exercise improve your writing?
• Listening	(1) How do listening and silence differ? (2) Enumerate the purposes of listening.	When is listening effective?	What do you need to improve your listening ability?
• Oral presentations	How important are oral presentations for professional accounting?	Which sense data (voice, appearance, fear, activities, etc.) are important to judge an oral presentation's effectiveness?	What did you learn from your actual oral presentation in class?
• Synthesizing	When and why do accountants need to synthesize? How do synthesizing and analyzing differ?	What new knowledge does synthesizing in an actual experience reveal?	How does synthesizing relate to integration of concepts? To globalization?
• Observing	What do auditors observe in an audit engagement?	When is observing essential ? Effective?	What are the key points of an exercise on observing?
• Comparing	Should accountants consider both differences *and* similarities when they compare?	What is a *valid* comparison?	How does this skill relate to the accounting convention of "comparability?"
• Classifying	Determine the purposes of classifying in a given exercise. What are the bases (criteria) for developing a classification?	Are the purpose(s) and criteria of the classification compatible?	Explain the purposes and criteria of classification of the "chart of accounts."

skepticism as an attitude, for example, a questioning mind and critical assessment of audit evidence including (1) sensitivity in selecting the nature and extent of documentation; and, (2) recognizing the need to corroborate management explanation or representations concerning material matters. Thus, the **reasoning** step of critical thinking directs students to seek and explain explicit and hidden assumptions in a given inquiry. The **evaluating** step centers on judging the validity and significance of these assumptions. In the process, students may discover the presence of self-serving assumptions, faulty logic, question-begging answers, circular argumentation, and ungrounded conclusions. This discovery is a key to learning as it both enhances the incremental volume of knowledge and injects into the mind a sense of self-accomplishment, an "intellectual high," that drives one to exert more effort and immersion into knowledge acquisition and utilization.

Table 3 explains how the above listed skills apply the three components of critical thinking to accounting education. While each listed skill may employ all of the three critical thinking components (reasoning, evaluating and self-reflecting) simultaneously to different degrees, we explain these components separately by examples for the sake of simplifying the presentation.

Applying Critical Thinking to Accounting Classes

To demonstrate the *effect* role critical thinking skills play in actual class settings, we briefly describe below a few experiments and views that some accounting educators applied or recommended.

The Goal in Managerial Accounting Courses

Goldratt's (1992) best-selling book, *The Goal*, provides a novel approach to attract managers' attention to the importance of critical thinking. Houston and Talbott (1993) report on their senior managerial accounting class applying Goldratt's principles of bottlenecks, inventory problems, accounting evils and continuous improvement--including (1) unneeded restructuring in the student library that added layers of administration and further distanced departments from the source of their operating budgets; (2) a freight shipper ignoring bottlenecks that caused some employees to work overtime while other ones sat idly when the conveyer belt broke down; and (3) a manufacturer using a just-in-time inventory technique whose supplier made a late delivery, thus causing the manufacturer unnecessary overtime costs.

Emphasis on Degree of Structure and Direction in Accounting Courses

Riordan and St. Pierre (1992) urge that accounting faculty members first develop students' critical thinking skills before improving their economic or accounting framework or even "number crunching" skills. They emphasize that rather than focus on *what* approach to use (procedural versus conceptual), accounting educators stress the *degree* to which the approach is structured and directed. For courses such as intermediate and cost accounting, faculty members should go beyond traditional lectures that encourage students to merely memorize content rather than think critically.

The Team Approach to Class Coverage

Hardy et al. (1993) summarize how Brigham Young University, adhering to the AECC's directive to restructure and otherwise improve accounting education, revised its accounting curriculum. Instead of taking a traditional sequence of courses (e.g., from accounting principles to auditing), students take a team-taught "course" four to five days per week in three-hour blocks each day. For example, students could learn the financial, managerial, tax, auditing and systems components of the cash collections cycle. To help students learn "how to learn," each of the 112 teaching plans that 11 professors developed for this course contained some critical thinking components—as well as written and oral communications, group work, and unstructured problem assignments. Therefore, topics taught in different functional area courses are brought together and compared rather than taught in isolation. For example, students learn how auditing, accounting information systems, cost accounting, finance, and legal environment, among other disciplines, view and treat the internal control system.

Role Playing in MBA Classes

Franz's (1989) MBA students play the roles of managers in their capacity of making strategic decisions and presenting them to student "corporate" boards of directors, while asking the rest of the class to critically analyze these presentations to help students learn "by doing" in vivid simulation exercises in accounting and other business areas.

Problems, Pitfalls, And Insights Into Applying Critical Thinking Skills

Involving students in learning and practicing critical thinking skills in class requires time and genuine efforts from instructors and students. Accordingly, some trade-off decisions become necessary regarding replacing some older topics and methods of teaching with new critical-thinking-based approach to learning. These tradeoffs include the following:

1. *Homework.* We select homework problems and exercises that require critical thinking skills to replace those assignments that motivate memorizing and mere number crunching without reasoning. Furthermore, instead of merely copying solu-

tions to assigned problems from the overhead projector screen, students receive copies of these solutions after giving them ample time to work them out. This helps them to participate in discussing the solutions and raise questions on areas of concern; however, as explained below, we must motivate the few students who do not attempt to do the assigned exercises before their coverage in class.

2. *Class Participation.* Not all students participate in class discussions for various reasons (e.g., shyness, lack of confidence, lack of interest, and fatigue). Since critical thinking skills in class demands students' active participation, we give students "bonus" points to their final grade (up to three-points per term) when they "add value" to class discussions. Participation can take such forms as asking critical questions, responding to the instructor's and fellow students' questions, synthesizing different concepts and practices, bringing in current news materials, or criticizing textbook presentations under discussion. We found that even a small percentage of the total grade assigned to class participation can motivate most students to participate in and out of class. To minimize the time taken to record such student participation, we use class-seating charts, which also help us memorize student names and add to their perception that we recognize all of them and expect their involvement.

3. *Incomplete Handouts.* While most students appreciate course handouts, some students may rely completely on these handouts without fully reading the course textbook and literature. To minimize this problem, we designed the handouts to include spaces for students to complete, which also allows us to ask questions that require students to think critically. For example, in cost accounting, instead of merely defining efficiency and effectiveness, we ask them how do they measure these attributes of a new multimillion dollar service program they were promoted to operate and manage; their job security, promotion and bonuses depend on how efficient and effective they perform. Gradually, the discussions lead to what to measure, how to measure, surrogates to measure, and the relative nature of efficiency and effectiveness criteria. Traditionally, most instructors do not distribute their class notes and summaries. By designing and sharing incomplete chapter outlines, we also discipline students to streamline their thinking to remain on track--rather than extending into irrelevant details and peripherals; this thought direction helps explain why we call accounting a *discipline*.

4. *Group Projects.* To help students develop teamwork skills, we assign group writing and computer projects and encourage them to work in teams on these assignments. However, we need to spend some class time going over the instructions, expected output, criteria for grading, and even helping some students, usually after class, with basic spreadsheet computing, designing a research paper, and locating references and manuals.

5. *Examinations.* We constantly remind students that understanding rather than mere memorizing is the key to excel on exams. We choose problems and essays that require critical thinking skills, for example, how CPAs' responsibility for fraud has changed with the recent court decisions and recent AICPA

pronouncements that Scantron multiple-choice questions taken from a text's test bank do not cover.

6. *Colleagues.* If two instructors teach different sections of the same course, the instructors should agree on common texts, syllabi, and examinations, which urge new and adjunct professors to use critical thinking skills.

Teaching based on learning critical thinking skills normally instigates time reallocation from research and service endeavors to teaching--a reallocation that could reduce faculty's merit pay. Fortunately, our programs adopt teaching portfolios to help assess the teaching component of faculty merit pay (Calderon and Green 1997, 223). Our faculty teaching evaluation committees generally look favorably on instructors' efforts on this new teaching methodology.

In summary, in order for students to learn the critical thinking skills described in this paper, accounting educators must (1) become critical thinkers themselves, and (2) possess and show in class genuine enthusiasm for learning via reasoning, evaluating, and self-reflecting.

SUMMARY AND CONCLUSIONS

Critical thinking plays two roles in accounting education. As a *cause*, many critics have found significant weaknesses in accounting education's qualitative and quantitative dimensions for many years. In the *qualitative* dimension, earlier accounting education focused on knowledge acquisition, technical procedures, semi-vocationalism, and passing the uniform CPA examination. The *quantitative* dimension was driven by three major factors: (1) the sheer size of authoritative pronouncements; (2) the fixed number of credit hours available for teaching; and, (3) educators feeling obligated to teach all technical aspects of accounting procedures. This concentrated critique of accounting education over many years has created the urge for change to embrace broader goals and to integrate knowledge.

As many studies and organizations recently have recommended significant changes to accounting curricula, critical thinking has gained a wide acceptance as a basic skill. This is the *effect* role critical thinking plays in accounting education. However, since critical thinking skills lack a generally accepted definition in accounting and they consist of complex concepts, this paper critically examines the concepts of "thinking," "criticality," and " skills" in order to help educators implement the recommended changes to their programs.

Thinking with goal-labeled "computation-level of thinking" has global structure and emphasizes the function of thinking (i.e., *what* the mind is doing). This is the appropriate thinking level for accounting education since the function of thinking forms the basis for accounting knowledge acquisition *and* utilization. Thinking *critically* contains three components (reasoning, evaluating and self-reflecting) that render negative, positive or neutral characteristics. In psychology, the term "skill" has received much attention. After reviewing some of these studies, we conclude that the "computational level of thinking" is consistent with "computational skills" defined as "the ability to execute an internal hierarchical organization" (Johnson-Laird 1991, 430). Some psychologists call this organization a "plan" which they define as "any hierarchical process in the organization that can control the order in which a sequence of operations is to be performed" (Miller et al. 1960, quoted in Johnson-Laird 1991, 431). Thus, in developing a skill for the computational-level of thinking in accounting education, the strategy is to focus on helping students recognize the hierarchical process needed in a thinking situation and learn how to plan and perform an efficient sequence of operations that leads to the goal of the thinking endeavor.

Finally, we show how several skills recommended for the new accounting education can apply the three components (reasoning, evaluating and self-reflecting) of critical thinking.

REFERENCES

American Accounting Association (AAA). 1967. Committee to Compile a Revised Statement of Educational Policy, R.J. Canning, Chair. A restatement of matters relating to educational policy. *The Accounting Review, Committee Report*, 50-121.

American Accounting Association (AAA). 1991. Accounting Education Change Commission (AECC). AECC urges decoupling of academic studies and professional accounting examination preparation: Issue Statement No. Two. *Issues in Accounting Education* (Fall): 313-314.

American Accounting Association (AAA). 1986. Committee on the Future Structure, Content, and Scope of Accounting Education (The Bedford Committee). Future accounting education: Preparing for the expanding profession. *Issues in Accounting Education* (Spring): 168-195.

American Institute of Certified Public Accountants. 1997. Statement on Auditing Standard (SAS) No. 82. *Consideration of Fraud in a Financial Statement Audit* (February).

Amernic, J.H. 1985. Teaching accounting in a vacuum. *CA Magazine* (May): 67-71.

Bayou, M., and E. Panitz. 1993. Definition and content of persuasion in accounting. *Journal of Applied Business Research* 9(3): 44-51.

Calderon, T.G., and B.P. Green. 1997. Use of multiple information types in assessing accounting faculty teaching performance. *Journal of Accounting Education* 15(2): 221-239.

Cassagio, J. 1997. How to integrate active learning skills into your course. *Colloquium on Change in Accounting Education* (November): 1-22.

Cunningham, B. 1996. How to restructure an accounting course to enhance creative and critical thinking. *Accounting Education* 1(1): 49-66.

Dellarosa, D. 1991. A history of thinking. Pp. 1-18 in *The Psychology of Human Thought*, edited by R.J. Sternberg and E.E. Smith. New York: Cambridge University Press.

Deppe, L.A., E.O. Sonderegger, J.D. Stice, D.C. Clark, and G.F. Streuling. 1991. Emerging competencies for the practice of accountancy. *Journal of Accounting Education* (Fall): 257-290.

Ericsson, K.A., and W.L. Oliver. 1991. Methodology for laboratory research on thinking: Task selection, collection of observations, and data analysis. Pp. 392-428 in *The Psychology of Human Thought*, edited by R.J. Sternberg and E.E. Smith. New York: Cambridge University Press.

Federation of Schools of Accountancy. 1997. *Critical Thinking Competencies Essential to Success in Public Accounting*. Report of the 1997 Educational Research Committee, C. P. Baril (Chair).

Finocchiaro, M.A. 1990. Critical thinking and thinking critically: Response to Siegel. *Philosophy of the Social Sciences* (December): 462-465.

Finocchiaro, M.A. 1989. Siegel on critical thinking. *Philosophy of the Social Sciences* 19: 483-492.

Franz, L. 1989. Integrating analysis and communication skills into the decision sciences curriculum. *Decision Sciences* 20: 830-843.

Gardner, H. 1985. *The Mind's New Science: A History of Cognitive Revolution.* New York: Basic.

Goldratt, E.M., and J. Cox. 1992. *The Goal.* New Haven, CT: North River Press.

Gronhaug, K., and G. Kaufmann 1988. Introduction. Pp. 1-10 in *Innovation: A Cross-disciplinary Perspective,* edited by K. Gronhaug and G. Kaufmann. Oxford: Oxford University Press.

Hardy, J.W., L.A. Deppe, and J.M. Smith. 1993. A curriculum for the 1990s and beyond. *Management Accounting* (September): 66.

Haugeland, J. 1985. *Artificial Intelligence: The Very Idea.* Cambridge, MA: Bradford/MIT Press.

Hofer, B.K., and P.R. Pintrich. 1997. The development of epistemological theories: Beliefs about knowledge and knowing and their relation to learning. *Review of Educational Research* 67(1): 88-140.

Houston, M., and J. Talbott. 1993. Critical thinking and *The Goal. Management Accounting* (December): 60.

Institute of Management Accountants (IMA). 1997. Drastic changes in management accounting, financial management make seven specific skills necessary for success. *News* (September): 1-4.

Johnson, H.T. 1992. *Relevance Regained: From Top-down Control to Bottom-up Management.* New York: Free Press.

Johnson, H., and R.S. Kaplan. 1987. *Relevance Lost: The Rise and Fall of Management Accounting.* Boston: Harvard Business School Press.

Johnson-Laird, P.N. 1991. A taxonomy of thinking. Pp. 429-457 in *The Psychology of Human Thought,* edited by R.J. Sternberg and E.E. Smith. New York: Cambridge University Press.

Kabanoff, B., and J.R. Rossiter. 1994. Recent developments in applied creativity. Pp. 283-324 in *International Review of Industrial and Organizational Psychology,* edited by C.L. Cooper and I.T. Robertson. Chichester: Wiley.

Keating, P.J., and S.F. Jablonsky. 1990. *Changing Roles of Financial Management: Getting Close to the Business.* Morristown, NJ: Financial Executive Research Foundation.

Keegan, D.P., and S.W. Portik. 1995. Accounting will survive the coming century, won't it? *Management Accounting* (December): 24-29.

Langenderfer, H.Q. 1987. Accounting education's history: A 100-year search for identity. *Journal of Accountancy* (May): 302-331.

Marr, D. 1982. *Vision: A Computational Investigation into the Human Representation and Processing of Visual Information.* New York: Freeman.

May, C., and G.S. May. 1996. *Effective Writing,* 4th ed. Engelwood Cliffs, NJ: Prentice-Hall.

The Merriam-Webster's English Dictionary. 1995. Springfield, MA: Merriam-Webster.

Miller, G.A., E. Galanter, and K. Pribram. 1960. *Plans and the Structure of Behavior.* New York: Holt, Rinehart & Winston.

Nelson, I.T. 1995. What's new about accounting education change? An historical perspective on the change movement. *Accounting Horizons* (December): 62-75.

Newell, A., and H.A. Simon. 1972. *Human Problem Solving.* Cambridge, MA: Harvard University Press.

Perspectives on education: Capabilities for success in the accounting profession. 1989. New York: Arthur Andersen & Co., Arthur Young, Coopers & Lybrand, Deloitte Haskins & Sells, Ernst & Whinney, Peat Marwick Main & Co., Price Waterhouse, and Touche Ross.

Previts G.J., and B.D. Merino. 1979. *A History of Accounting in America: An Historical Interpretation of the Cultural Significance of Accounting.* New York: Ronald Press.

Reckers, P.M.J. 1995. Know thy customers. Pp. 29-35 in *Change in Accounting Education: A Research Blueprint,* edited by C. Baril. Federation of Schools of Accountancy.

Riordan, M.P., and E.K. St. Pierre. 1992. The development of critical thinking. *Management Accounting* (February): 63.

Rottenberg, A.T. 1985. *Elements of argument: A text and reader.* New York: St. Martin's Press.

Siegel, G., and C. S. Kulesza. 1995. Encouraging change in accounting education. *Management Accounting* (May): 19-23.

Siegel, H. 1990. Must thinking be critical to be critical thinking? Reply to Finocchiaro. *Philosophy of the Social Sciences* (December): 453-461.

Siegel, H. 1988. *Educating Reason: Rationality, Critical Thinking, and Education.* London: Routledge.

Svetcov, D. 1995. Offering custom-made MBAs. *The Chronicle of Higher Education* (December): A17-A18.

U.S. News and World Report. 1993. The M.B.A. gets real. (May 22): 54-57.

Williams, D.Z. 1993. Reforming accounting education. *Journal of Accountancy* (August): 76-82.

Zeff, S.A. 1989. Does accounting belong in the university curriculum? *Issues in Accounting Education* (Spring): 1-3.

FOSTERING CRITICAL THINKING IN ACCOUNTING EDUCATION:
IMPLICATIONS OF ANALYTICAL PROCEDURES RESEARCH

James L. Bierstaker, Jean C. Bedard, and Stanley F. Biggs

ABSTRACT

This paper extends the literature on critical thinking in accounting education by drawing on our research results, which have identified specific decision process problems that auditors encounter when conducting analytical procedures. The auditors' difficulties in performing the required financial statement analysis illustrate the need to improve accounting students' critical thinking skills. Accounting students who are accustomed to highly structured problems in the traditional curriculum may be unprepared for the complexity of problem solving in practice. Based on the research findings, we present a framework for teaching critical thinking, which faculty can use to address the identified difficulties in financial statement analysis. Further, we provide two analytical procedures cases (one relatively simple, the other more complex) that can be used in auditing courses. The framework provides specific guidance to assist faculty using these and other financial statement analysis case situations to foster their students' critical thinking.

Advances in Accounting Education, Volume 2, pages 21-36.

INTRODUCTION

In 1990, the Accounting Education Change Commission called for educators to develop ways to enhance accounting students' critical thinking skills. Since that time, several articles have appeared in the accounting literature that provide support for faculty who wish to adapt their pedagogical approach in ways that may enhance students' critical thinking (e.g., Cunningham 1996, 55-62; Kimmel 1995, 303-307; Gabriel and Hirsch 1992, 243-245). This paper extends that line of literature by presenting a framework derived from a series of studies on auditors' decision processes (Bedard and Biggs 1991, 632-636; Bierstaker, Bedard, and Biggs 1999, 19-21; Bedard, Biggs, and Maroney 1998, 9). These studies identify specific steps in critical thinking processes that enhanced or inhibited practicing auditors' performance of analytical procedures. Thus, the framework developed from decision process evidence in the professional context of auditing both supports and provides specific direction to AECC initiatives on critical thinking. Further, we provide the case materials used in the research (New England Electronics) along with a preparatory case (Royal Company), and show how we have used these cases in the classroom to improve our students' approach to complex situations involving financial statement analysis.

Use of complex, relatively unstructured problems and cases in the classroom has been recommended as a means to enhance critical thinking skills (AAA 1986, 185-186). These recommendations are based on a lack of fit between the changing demands on accounting professionals and the modes of training traditionally used in most accounting curricula (Bell et al. 1997, x). The problems that students encounter in the traditional accounting classroom are often fairly straightforward applications of accounting or auditing standards. However, real-world accounting problems are more complex. For example, auditors must weave their way through a maze of financial and nonfinancial information, attempting to determine what information is relevant. They then must integrate relevant information into a meaningful pattern. They may generate, evaluate, and discard many potential solutions before finally selecting one. The process of exploring a problem situation to arrive at a justifiable conclusion that integrates all available information is called critical thinking (Kimmel 1995, 303-307; Kurfiss 1988, 42).[1] The premise of the AECC and the accounting profession, expressed in the above cited documents, is that practice in solving unstructured problems can lead to development of critical thinking skills, improving performance in the complex problems students will encounter in their careers.

Articles such as Kimmel (1995, 303-307) and Cunningham (1996, 55-62) provide recommendations to faculty in implementing curricula aimed at improving students' critical thinking skills. This paper builds on those articles by offering evidence of how auditors approach solving complex problems, and using that evidence to reinforce and extend prior recommendations. This evidence comes from a series of studies on auditors' use of analytical procedures to identify a particular error in a client's unaudited financial statements. Because we used a "think aloud" research method to trace decision processes, we were able to identify a series of decision process steps used by auditors who correctly solved the case, and some specific difficulties encountered by those who did not. These findings are relevant to accounting education because many of the difficulties we identify can be addressed through use of unstructured problems in the classroom. We have used the New England Electronics case and other cases to illustrate to our students a framework for critical thinking in analytical procedures, as well as other auditing tasks.

Although the setting of the studies was audit analytical procedures, the task shares features with other complex problems in accounting and auditing, suggesting that the benefits we discuss may be applicable to other settings. First, good performance in the task involves going beyond the accounts provided to compute other information using knowledge of financial statement relationships. Thus, this task is similar to many applications of financial statement analysis, including investment and credit decisions. Second, a number of solutions to the task might seem plausible at first, especially if auditors only examine some of the available information. However, a thorough evaluation shows that many solutions do not fit all the task information. Thus, as with many complex problems, the auditor must develop an initial set of possible solutions and then choose among them.

We begin by briefly describing audit analytical procedures and the methods used to capture auditors' decision steps. We then summarize results of the research series, draw implications for accounting education, and discuss how we have used the cases presented in the Appendices to improve our students' critical thinking skills.

DESCRIPTION OF THE RESEARCH AND IMPLICATIONS FOR CRITICAL THINKING

Context: Analytical Procedures in Auditing

Auditors use analytical procedures (AP) to determine if the client's unaudited account balances are different from the amounts that could be expected based on past financial performance, industry trends, the overall economy, etc. If AP provide a good signal of which accounts might be in error, the auditor can plan further audit tests to detect where restatement might be needed. In order to direct audit testing efficiently, the auditor needs some idea of what might have caused an unexpected discrepancy. For instance, if (s)he finds unexpected discrepancies in inventory, are they caused by poor cutoff procedures? Obsolescence? Changes in materials pricing? Good performance of AP will result if the auditor is able to consider a number of possible explanations and select those that should guide further audit tests. Audit firms are relying increasingly on AP as an efficient form of testing. However, because there are many possible explanations for unexpected account discrepancies (including economic conditions, industry trends, errors and fraud), AP can be complex and difficult.

Methods Used to Conduct the Research

Professional auditors participating in the studies responded to a case situation (New England Electronics, presented in Appendix B) in which an audit client is considering an initial public offering of stock in the coming year. In this case, we seeded a material overhead allocation error into a set of financial statement information (the baseline for comparison, termed "projected" information), creating the "unaudited" accounts and ratios. Specifically, the error involves improper capitalization of selling, general and administrative expenses (SG&A) into the inventory pool, which the client allocated to inventory and cost of sales. Material discrepancies between the projected and unaudited information include income and inventory that are above the projected amount, yet the gross margin is below. The auditor needs to consider the most likely causes of the discrepancies and develop an audit plan to investigate whether an error or irregularity has occurred. The case also contains a suggested explanation from the client, related to a large year-end raw materials purchase that does not adequately explain discrepancies. Auditors' decision processes were captured on audio tape as they performed the task while "thinking aloud." Thus, a detailed trace of their progress is available through all stages of the process from acquiring information to proposal of possible explanations for discrepancies.

Bedard and Biggs (1991, 632-636) described a four-step process used by auditors who performed well; that is, those who hypothesized an accounting error that explaining all unexpected discrepancies. This study also used decision process data to identify specific difficulties that *inhibited* good performance in each of the stages. A notable finding was that unsuccessful participants appeared to focus on a single type of solution that did not explain all the discrepancies found in the case and had difficulty thinking of others.

Following up that study, Bierstaker et al. (1999, 3-12) investigated why auditors become fixated on a certain solution, and how this fixation can be overcome. This study adapted and expanded upon a method from psychology (Kaplan and Simon 1990, 384) to explore how, by accessing different mental images or "representations" of a problem, a problem solver can consider new ideas and solutions. Using this method, we gave auditors helpful prompts to encourage them to shift their problem representations. The prompts we developed for the New England Electronics case are similar to suggestions that auditors might receive from co-workers (e.g., "this is a manufacturing firm with both product and period costs"). The study explored factors that help or hinder changes in problem representations in response to the prompts.

The third study (Bedard et al. 1998, 12-20) analyzed decision processes of auditors working in groups, to explore how auditors interact with other members of the audit team during planning. In business, we often assume that working in groups will enhance performance. However, research in psychology shows that interpersonal interaction can provide gain *or* loss relative to individual performance, depending on the nature of the task and the ways in which the group interacts. This study analyzed sources of group process gain, including effects of knowledge pooling (the contribution of ideas from multiple individuals) and error correction. Sources of group process loss include dominance by a group member and interruptions, both of which might prevent other members from contributing or fully defending their ideas.

Findings of the Studies and Related Suggestions for Improving Students' Critical Thinking

The Bedard and Biggs (1991, 632-636) process model suggests a framework for analyzing how auditors approach complex problems, where they encounter difficulty, and possible reasons for those difficulties.[2] Thus, the studies cited above provide evidence useful to faculty who are training future auditors in critical thinking. Table 1 summarizes the steps in this framework, the reasons for difficulties auditors encounter at each step, and specific suggestions for enhancing critical thinking in the classroom developed from the findings.

Step 1: Information Acquisition

To identify the case's underlying error in overhead allocation, auditors needed to identify four important discrepancies: inventory is overstated, sales are not affected, net income is overstated, and gross margin is understated. Verbal protocol analysis revealed that a few auditors did not acquire all of the relevant information, for several interrelated reasons. First, some started by focusing on single, easily identified cues, particularly the inventory overstatement. They then began generating possible solutions. While inventory is important, auditors also must consider other accounts in order to do a thorough financial statement analysis. Students may exhibit similar behavior in the classroom. To counteract it, instructors should encourage students to examine all case information before "jumping to conclusions."

Second, it may have been easier to acquire information that was readily available and ignore or delay acquiring information that would require a more thorough investigation. For example, although auditors could have used the financial ratios and balances given to calculate SG&A expense, several of them chose to focus mainly on the account balances presented. This suggests that students' critical thinking skills could be improved by incorporating problems that require them to search for and derive certain important pieces of information, which are

Table 1.   Steps in the Critical Thinking Framework, Reasons for
Difficulty and Suggestions for Improvement

Steps	Reasons for Difficulty Among Practicing Auditors	Guidance to Improve Students' Critical Thinking Skills

1. Information Acquisition
 	
Ideal: Acquire all relevant information.
	• Began generating solutions before all relevant information is acquired.
	• Did not derive information beyond that presented.
		• Before "jumping to conclusions":
		  • Consider all case information.
		  • Derive any important ratios or balances that are missing.

2. Pattern Recognition **Ideal:** Organize relevant information into a pattern to be explained.	• Focused on explaining a single account. • Not used to looking for patterns among accounts.	• Think about relationships within the financial and nonfinancial information.
3. Solution Generation **Ideal:** Generate solutions that may fit the pattern.	• Relied on a suggested solution instead of generating a new one. • Current understanding of the problem limited possible solutions, and changing that understanding was difficult.	• "Brainstorm" a variety of ideas before settling on one. • Try to ignore suggested solutions until you have developed your own ideas. • Try to think of the problem from more than one perspective. Are there different *kinds* of possible answers?
4. Solution Evaluation **Ideal:** Test possible solutions against the data and identify the best answer.	• Tested solutions using an incomplete set of accounts, focusing on those consistent with a solution being considered. • Continued to test previous solution(s) rather than generating new ones ("wheel-spinning").	• Now that you have a good set of possible answers, be sure to give each a fair test. • Don't just consider why an answer might be right, but why it might be *wrong*. • If your proposed solutions are not correct, make the effort to go back to the beginning and re-examine your first three steps.

more difficult to acquire. This is consistent with the suggestion of Huffman et al.
(1991, xx-xxi) that an important component of critical thinking is the ability to
analyze data for value and content.

Step 2: Pattern Recognition

The seeded error can be identified from the pattern of changes in the four important cues. Of those auditors who acquired all four cues, some focused on one
account at a time, not considering the cues as a pattern. Those auditors who expe-

rienced pattern recognition problems did not integrate balance sheet and income statement accounts. Bedard and Biggs' (1991, 633-635) finding that managers outperformed seniors may be related to managers having more experience in integrating accounts to achieve an overall understanding of the client's financial status. This suggests that accounting students may obtain a more global perspective with practice incorporating information from several sources (multiple financial statements, management's discussion and analysis (MD&A), industry information, etc.), and identifying relationships among them (see also AECC 1990, 311). Moreover, students may need repeated exposure to problems in which recognizing patterns is important, so they develop skills and problem solving habits that stress combinations of information rather than one-cue-at-a-time processing.

Step 3: Solution Generation

Some auditors whose verbal protocols showed that they understood the pattern of cues were not able to suggest an accounting error that fit that pattern. Those who had difficulty at this stage often focused on one type of accounting error, usually either recording or cutoff errors. Several auditors who had difficulty at this stage focused their proposed solutions around client management's explanation, even when they previously had rejected solutions based on that explanation as inconsistent with the evidence. Moreover, some auditors indicated that their next strategy would be to call the client for further ideas, even when solutions derived from that explanation did not fit. Field research indicates that client management is a common source of explanations for discrepancies in practice (Hirst and Koonce 1996, 462). Although obtaining explanations from the client may seem easier than generating other possibilities, auditing standards indicate that the client's assertions are not independent evidence, and should be objectively tested before being relied upon.

This behavior is reminiscent of students' reliance on their professors or solutions manuals to supply them with the correct answer. One way that accounting educators can address these difficulties is by requiring that students develop their own solutions prior to consulting an answer key. Information acquisition and integration problems resulting from the presence of suggested solutions also may account for students' difficulties with multiple choice questions. A focus on testing one or more of the suggested answers (working backward toward the information provided) rather than on acquiring information to move forward toward a solution, is less likely to result in good solutions (Bedard and Biggs 1991, 628-639).

Once auditors began proposing solutions, it was often difficult to change their initial understanding of the situation. Because possible solutions flow from the mental representation of a problem, some auditors had difficulty thinking of additional ideas once they found that their initial solution was incorrect. In the classroom, faculty can address this difficulty by encouraging students to consider a problem from different perspectives. This approach can be as simple as asking students who seem to be "stuck" to try another line of reasoning. Faculty also can illustrate the power of different perspectives by assigning students to adopt the roles of different parties in a case (see also, Cunningham 1996, 59). Another method that is adaptable to financial statement analysis problems is first to give groups of students different information and then have them come together to compare results. For example, some could start with financial information, while others examine nonfinancial information (such as the MD&A).

Step 4: Solution Evaluation

Bierstaker et al. (1999, 31-33) identified two solution evaluation tendencies that inhibited auditors from changing their understanding of the problem. One is "confirmation bias:" some auditors focused on cues that would support their proposed solutions or made errors while testing proposed solutions so the solutions appeared to fit the cues. Another is "wheel-spinning" or repeatedly looping back to re-test solutions rejected previously. Both confirmation bias and wheel-spinning are related to the tendency to "explain away" evidence that does not fit an explanation in mind (Schank 1986, 27-30). One way to counteract this tendency is to ask students to cite evidence both *for* and *against* particular answers.[3] After having done so, the instructor can inform students of the natural human tendency, found in general psychology research and within the auditing context, to prefer to examine information that supports a proposed answer over information that might refute it. This can lead to a productive discussion of the role of skepticism in auditing.

Group Interaction

When Bedard et al. (1998, 12-20) compared interacting groups of three audit seniors to individuals working on the New England Electronics case, they found that groups were better than individuals at *considering* a good solution. However, groups were not necessarily better overall at *evaluating* solutions. Process gain from group interaction resulted from knowledge pooling and error correction, which were important in helping groups generate a large number of possible solutions. The relative amount of interruption by group members affected their ability to select the right solution from that set. These results suggest that groups are productive in generating solutions (i.e., brainstorming), but that *how* the group interacts is very important in obtaining a favorable task outcome. These findings may benefit faculty in assigning group work in class. Based on the research results, student groups might benefit from some instruction on the *quality* of interaction, particularly in allowing expression of complete ideas from all members. Our research also suggests that groups are better at generating ideas than at evaluating them. Thus, faculty can break up work on cases into generation and evaluation components. First, the group can generate ideas while audio taping or taking careful notes. Next, individuals can evaluate those ideas, and then the group can come to a joint solution.

APPLYING THE FRAMEWORK: TWO CASES FOR TEACHING CRITICAL THINKING

In this section, we present two cases we have used to develop our students' critical thinking skills: Royal Company and New England Electronics. Regarding each, we discuss the nature of its underlying financial statement error, important nonfinancial information, and some of our experiences using the case. We have found that students often are surprised to learn that problems in the accounts can be identified using financial data, which may explain some of their initial difficulties with solving these cases. This observation underscores the importance of repeatedly exposing students to financial statement analysis in the classroom. In

Table 2. Applying the Critical Thinking Framework in Two Analytical Procedures Cases

Steps	Royal Company	New England Electronics
1. Information Acquisition	• Current ratio is overstated.	• Inventory is overstated.
	• Current assets are at	• Net income is overstated.
Ideal: Acquire all relevant	expectations.	• Gross margin is understated.
information.	• Long-term liabilities over-stated.	• Sales are at expectations.
	• Net income is an expected.	
2. Pattern Recognition	• Since the current ratio is overstated and current assets	• Inventory is higher than expected.
Ideal: Organize relevant	are at expectations, current	• Sales are as expected and
information into a	liabilities must be	gross margin is understated,
single pattern to be	understated.	so cost of goods sold must
explained.		also be overstated.
		• Gross margin is understated and income is overstated, so SG&A expenses must be understated.
3. Solution Generation	• Problems with current liabilities	• Problems involving both inventory (balance sheet) and
Ideal: Generate solutions that		operating expenses (income
fit the pattern.		statement).
4. Solution Evaluation	• Misclassification of long-term debt as short-term.	• Improper allocation of overhead from period to
Ideal: Test possible solutions		product cost.
and identify the correct		
one.		

addition to use of the cases, we discuss the importance of skill in identifying potential problems in financial statements for a variety of accounting careers, such as internal and external auditing, banking, managerial accounting, financial planning, and regulation. In contrast to a more generic approach to using case studies (e.g., answering a series of discussion questions), we teach the cases using the four-step framework described in Table 1. Important features of each case are summarized in Table 2.

The Royal Company Case

The Royal Company case (see Appendix A) contains a single problem that causes a fairly straightforward pattern of discrepancies in the financial ratios and account balances. The case involves an audit client that has failed to reclassify a portion of long-term debt as short-term. This error results in the following pattern of discrepancies between the client's financial ratios, account balances and the auditor's expectations: the current ratio is overstated, the current assets balance is the same as expected, the long-term liabilities balance is overstated, and net income is as expected. Nonfinancial information in the case reveals that the client recently received a loan with a debt covenant that requires the current ratio to be maintained at a certain level. Thus, the client has a motivation to overstate the current ratio if they are at risk of violating the terms of the loan agreement.

This case is effective for teaching students using the critical thinking framework outlined above. At the *information acquisition* stage, it illustrates the importance of a thorough information search prior to generating potential solutions, including how to derive important pieces of information. For example, students can use the current ratio and current asset balance to derive the current liabilities balance, which is understated. The current liabilities balance is a critical piece of information that students must calculate in order to identify the underlying problem correctly. At the *pattern recognition* stage, students must recognize the relationship between current assets and the current ratio in order to realize there is a problem with current liabilities. In our experience, students tend to focus initially on current assets, although the information given suggests that there is no reason to expect a problem in that area. We explain to the students that since the current asset balance is as expected, proposed solutions relating to problems with current assets can be rejected. This illustration leads them to begin to generate and evaluate potential solutions with current liabilities. At this point, students often propose that the client has failed to record a current liability. Although this proposal is consistent with the discrepancy in current liabilities, it does not explain the overstated long-term liabilities balance. When students see the understated current liabilities, overstated long-term liabilities, and consistent income as a pattern, they can identify a debt misclassification problem as a potential solution.

By beginning with a case that contains a fairly simple account pattern with only a few important pieces of information to acquire, students begin learning the framework for solving unstructured problems successfully. In addition, we use this case to discuss why current liabilities are an important area of audit focus, and that debt misclassification problems occur relatively frequently in practice (e.g., Coakley and Loebbecke 1985, 237). Another advantage of this case is that it can lead to discussion of reasons why such an error might have occurred. One possibility is that a portion of a long-term loan may come due without the client realizing it. Another possibility, suggested by the case's nonfinancial information, is that the client is aware of the upcoming due date but may want to avoid reclassification due to concern about debt covenant violation. Discussion can center on the different audit implications of these two possibilities.

The New England Electronics Case

The New England Electronics case (see Appendix B) is more challenging than is the Royal Company case, as the pattern of discrepancies is more complex. This case involves an audit client that has allocated excess SG&A expenses to the inventory and cost of goods sold balances. As with Royal Company, students must derive a critical piece of information (i.e., the amount of SG&A expense) to complete the information acquisition step. To recognize the pattern of discrepancies, they must link the overstated inventory and overstated net income with understated gross margin (and sales at the expected amount). Unlike the Royal Company problem that was confined to balance sheet accounts, students in this case must acquire both balance sheet and income statement information to identify the problem. In addition, they must draw on their managerial accounting knowledge of product and period costs. Pertinent non-financial information in the case includes the client's initial public offering of stock, providing a motivation to overstate income.

Similar to Royal Company, New England Electronics is useful for emphasizing the importance of a thorough analysis of the financial and nonfinancial information, including information that students must derive. In addition, we emphasize the importance of seeing the pattern of discrepancies before generating solutions. We have found that students' problem-solving strategies in this case are similar to the auditors we have studied in that they frequently focus on the overstated inventory and ignore other pertinent information such as overstated net income. In particular, students often suggest that there may be a problem with the inventory costing method. When they test this explanation, they find that it does not account for all of the discrepancies in the case, that is, both inventory being overstated and gross margin being understated. Many of the other solutions students commonly propose, such as a cutoff problem with inventory or a recording error, also can be tested and rejected by considering all of the information in the case. We impress upon students that in this case, as in their future careers, it may be necessary to test carefully and reject a variety of potential solutions until the true cause of discrepancies can be identified.

As noted above, this case also contains a client suggestion that the source of the discrepancies is a large purchase of raw materials at favorable prices near the end of the year. This explanation does not account for all discrepancies. We have found that helping students to see why it should be rejected is an effective way of demonstrating the importance of careful evaluation of proposed solutions. Moreover, we have used this explanation to impress upon our students the importance of professional skepticism. Auditors may compromise audit effectiveness by overreliance on assertions from clients, due to their inherent conflict of interest as the preparers of the financial statements. We direct students' attention to the client's upcoming initial public offering as a potential motivation for overstating net income, leading to a discussion of real companies that have been subject to Securities and Exchange Commission enforcement actions for overstating income.

Introducing the Cases into the Curriculum

Faculty can assign either case as a homework problem, an in-class assignment, or a group project. In addition, faculty can also use mini-cases on exams to emphasize the importance of critical thinking, and to reinforce students' memories of patterns of discrepancies they have encountered in the case studies discussed in class. Instructors should be aware that accounting students may be accustomed to highly structured problems where they use all the given information to solve the problem correctly. Students may become frustrated by less structured problems or cases that require them to determine what information is relevant or irrelevant for achieving the correct solution (Knechel 1992, 211-212; Craig and Amernic 1994, 41). In addition, using unstructured problems may be very time consuming, and make it difficult for the instructor to cover all of the course topics (Libby 1991, 196). Instructors may wish to introduce unstructured problem-solving in the classroom gradually, beginning the semester with quantitative problems and building toward more comprehensive cases with larger information sets and more detailed analysis later in the course. In this way, students can become acclimated to the new instructional techniques. Alternatively, instructors may wish to change their teaching style gradually by adding only a single new instructional technique each semester, in order to assess the effectiveness of each. Thus, the amount of new material covered in any one semester will not become overwhelming (Cunningham 1996, 63).

CONCLUSIONS

The purpose of this paper is to inform accounting educators of evidence we have derived from study of decision processes of practicing auditors, which has implications for accounting pedagogy. This evidence suggests that auditors have difficulty evaluating and integrating information from multiple sources. Particularly, auditors appeared to have difficulty identifying problems in financial statements, and were sometimes unable to change the way they were thinking about the problem. In addition, some auditors relied too heavily on suggestions from client management.

These results are consistent with difficulties in the critical thinking process. Based on our research and prior literature, we develop a framework that faculty can use for financial statement analysis cases and present two cases for use in this manner. The professional auditors' decision processes, and our own experience using the cases in the classroom, suggest points in the process at which students are likely to experience difficulty. Thus, this paper contributes to the accounting education literature by drawing on research evidence. While this approach has potential, use of complex, unstructured cases in the classroom is only one step toward enhancing professional accountants' critical thinking. Continued practice during job training, as well as relevant work experience, also are likely to be important factors in the development of this skill. Accounting educators should work toward developing techniques to assess progress in improving critical thinking, so that they can evaluate the effectiveness of various approaches.

APPENDIX A

Royal Company

Some ratios and balances for Royal Company, an audit client, are presented on the following page. You can assume that the projections are reliable, and are based on information that includes financial results from the first three quarters of the current year (19X1), trends from past years, and industry data.

Royal Company recently received a loan from First Bank. Terms of the loan require that Royal Company maintain a current ratio of 2.0 or better. In past audits, the audit team has chosen not to rely on internal controls because there are too few office personnel for proper segregation of duties. However, you feel that management is competent and trustworthy.

You may assume that any discrepancies between the reliable projected numbers and the client's unaudited numbers are material and are caused by a **single** error or irregularity. Your task is to examine the financial information, and list the simple accounting errors or irregularities that are most likely to be the cause of the discrepancies between the projections and unaudited figures. For each accounting error identified, give the corresponding debit and credit that you would recommend as an adjusting entry. Please be as specific as possible in describing the errors or irregularities that may have occurred, and list them in order of likelihood.

Table A1.   Royal Company Financial Information

Ratios	19X1 Projected	19X1 Unaudited
Current Ratio	1.68	2.00
Quick Ratio	0.73	0.90
Gross Margin Percentage	27.0%	27.0%
Income Before Taxes as a Percent of Sales	12.0%	12.0%
Inventory Turnover	10.7	10.7
Receivable Turnover (sales/ ending receivables)	24.2	24.2
Balances		
Inventory (FIFO), 12/31/X1	$320,000	$320,000
Current Assets, 12/31/X1	$580,000	$580,000
Long-term Liabilities, 19X1	$600,000	$655,000
Net Income, 19X1	$303,600	$303,600
Sales, 19X1	$4,400,000	$4,400,000

APPENDIX B

New England Electronics

You recently were assigned as the manager in charge of the audit of New England Electronics, Inc., a manufacturer of control systems. New England began operations in 19X1, and your firm has provided audit services since that time. The company is a closely held corporation that has been trying to break into a market dominated by overseas competition. Earnings have been growing at about 10% per year since 19X1, and in 19X4 had reached $450,000. Encouraged by this success, management is considering a public offering of stock next year.

The senior has approached you with some differences found in the comparison of unaudited current year balances and ratios to projections developed during preliminary audit work. The projections are based on results of the first three quarters, past audited balances, and industry trends. Having reviewed the projections, you are confident they are based on sound assumptions and information. Also, you have no reason to believe that conditions have changed that would affect the projections. In past audits, the audit team has chosen not to rely on internal controls because the small number of office personnel prohibits segregation of duties.

The senior is particularly concerned about the increase in income. Comparisons of liquidity (quick and current ratios) and profitability (gross margin and income to sales) also seem different, as well as inventory turnover. When the senior requested an explanation from management, client management asserted that discrepancies seen in the pattern of ratios are due to a large end-of-year purchase of raw materials, which was made to take advantage of favorable prices.

Your task is to assess the financial information shown on the next page, and to assist the senior in identifying a single accounting error that is most likely to be the cause of the discrepancies between the projected and actual figures. Once you have analyzed the financial information, please be as specific as possible in describing the most likely error that may have occurred. Assume that any discrepancies between the projected and actual numbers are material and are caused by a single error.

Table B1. New England Electronics Financial Information

Ratios	19X5 Projected	19X5 Unaudited
Current Ratio	1.50	1.52
Quick Ratio	1.00	0.98
Gross Margin Percentage	28.5%	27.8%
Income Before Taxes as a Percent of Sales	7.5%	10.0%
Inventory Turnover	2.85	2.62
Receivable Turnover (Sales/ Ending Receivables)	3.00	3.00
Balances		
Inventory (FIFO), 12/31/X5	$1,500,000	$1,650,000
Current Assets, 12/31/X5	$4,496,000	$4,646,000
Net Income, 19X5	$360,000	$270,000
Sales, 19X5	$6,000,000	$6,000,000

ACKNOWLEDGMENTS

The authors would like to thank the editor, Bill Schwartz, two anonymous reviewers, Bill Felix and Jim Rebele for their helpful comments. We also thank the participants of the 1997 Midyear Auditing Section Conference and the 1997 Northeast Regional Conference.

NOTES

1. We follow Kimmel's (1995, 303-307) approach by considering the problem-solving process from information integration and discovery through the final decision as "critical thinking." Cunningham (1996, 51) separates stages of this process into "creative thinking," defined as generation of possible solutions to a problem, and "critical thinking," evaluation of those solutions prior to choice.

2. This model, derived from auditors' decision processes, maps well into critical thinking models such as that of Kurfiss (1988, 42), who suggests that critical thinking involves integration of information, search for patterns, forumulaiton of hypotheses and justification of conclusions.

3. Research has shown that auditors do not develop reasons why their answers might not be right unless specifically asked to do so. When they do, this process affects their judgments about the likelihood of particular causes of financial statement discrepancies (Koonce 1992, 74).

REFERENCES

Accounting Education Change Commission. 1990. Objectives of Education for Accountants: Position Statement No. 1. *Issues in Accounting Education* (Fall): 307-12.

American Accounting Association Committee on the Future Structure, Content, and Scope of Accounting Education (Bedford Committee). 1986. Future accounting education: Preparing for the expanding profession. *Issues in Accounting Education* 1: 168-95.

Bedard, J., and S. Biggs. 1991. Processes of pattern recognition and hypothesis generation in analytical review. *The Accounting Review* 66(July): 622-42.

Bedard, J., S. Biggs, and J. Maroney. 1998. Sources of process gain and loss from group interaction in performance of analytical procedures. *Behavioral Research in Accounting* (Supplement): 1-27.

Bell, Y., F. Mars, I. Solomon, and H. Thomas. 1997. *Auditing Organizations through A Strategic-Systems Lens*. Montvale, NJ: KPMG Peat Marwick LLP.

Bierstaker, J., J. Bedard, and S. Biggs. 1999. The effect of problem representation shifts on auditor performance in analytical procedures. *Auditing: A Journal of Practice and Theory* (Spring): 18-36.

Coakley, J.R., and J.K. Loebbecke. 1985. The expectation of accounting errors in medium-sized manufacturing firms. Pp. 199-245 in *Advances in Accounting*, Vol. 2, edited by B.N. Schwartz. Greenwich, CT: JAI Press.

Craig, R., and J. Amernic. 1994. Role-playing in a conflict resolution setting: Description and some implications for accounting. *Issues in Accounting Education* 9(Spring): 28-44.

Cunningham, B.M. 1996. How to restructure an accounting course to enhance creative and critical thinking. *Accounting Education: A Journal of Theory, Practice and Research* 1(1): 49-66.

Ericsson, K., and H. Simon. 1993. *Protocol Analysis: Verbal Reports as Data*, rev. ed. Cambridge MA: The MIT Press.

Gabriel, S., and M. Hirsch. 1992. Critical thinking and communication skills: Integration and implementation issues. *Journal of Accounting Education* 10(2): 243-270.

Hirst, E., and L. Koonce. 1996. Audit analytical procedures: A field investigation. *Contemporary Accounting Research* (Fall): 457-485.

Huffman, K., M. Vernoy, B. Williams, and J. Vernoy. 1991. *Psychology in Action*. New York: John Wiley & Sons.

Kaplan, C., and H. Simon. 1990. In search of insight. *Cognitive Psychology* 22: 374-419.

Kimmel, P. 1995. A framework for incorporating critical thinking into accounting education. *Journal of Accounting Education* 13(3): 299-318.

Knechel, W.R. 1992. Using the case method in accounting instruction. *Issues in Accounting Education* (Fall): 205-17.

Koonce, L. 1992. Explanation and counterexplanation during audit analytical review. *The Accounting Review* 67(1): 59-76.

Koonce, L. 1993. A cognitive characterization of analytical review. *Auditing: A Journal of Practice and Theory* (Supplement): 57-76.

Kurfiss, J.G. 1988. *Critical thinking: theory, research, practice, and possibilities*. Washington, DC: Association for the Study of Higher Education.

Libby, P.A. 1991. Barriers to using cases in accounting education. *Issues in Accounting Education* 6: 193-213.

Libby, R. 1985. Availability and the generation of hypotheses in analytical review. *Journal of Accounting Research* (Autumn): 648-67.

Maroney, J., and J.C. Bedard. 1997. Auditors' treatment of inconsistent evidence during hypothesis evaluation. *International Journal of Auditing* (October): 187-204..

Schank, R. 1986. *Explanation Patterns*. Hillsdale, NJ: Lawrence Erlbaum.

INTEGRATING LEARNING STRATEGIES IN ACCOUNTING COURSES

Barbara J. Eide

ABSTRACT

Influential professional bodies (AAA 1986; AECC 1990, 1992; Arthur Andersen & Co. et al. 1989; CPA Vision 1998) have stated that accounting education's primary classroom objective is to help students learn how to learn. Accounting educators are encouraged to direct students' attention to the learning process and to help them learn how to learn. Learning to learn can be conceptualized as a process of acquiring, understanding, and using a variety of learning strategies to help one learn more effectively. This paper describes the learning process, defines learning strategies, and suggests ways in which accounting instructors can integrate several of these strategies in accounting courses to investigate whether students' learning is enhanced. The focus shifts from using learning strategies as an instructional tool to empowering students, through learning strategy use, to become independent learners.

Advances in Accounting Education, Volume 2, pages 37-55.
ISBN: 0-7623-0515-0

INTRODUCTION

In recent years, the Accounting Education Change Commission[1] (AECC 1990, 1992), the American Accounting Association (AAA 1986), the then Big-Eight managing partners (Arthur Andersen & Co. et al. 1989), and the Institute of Management Accountants (Siegel and Sorensen 1994) have urged that significant changes take place in the design and delivery of accounting education. Accounting programs are not keeping up with the dynamic, complex, expanding, and constantly changing accounting profession that students are being prepared to enter (AECC 1990). The Bedford Committee (AAA 1986, 172) wrote that "professional accounting education...has remained substantially the same over the past 50 years." As public evidence of the need to change accounting educational processes, the Committee cited complaints that "accounting graduates do not know how to communicate, cannot reason logically, and have limited problem-solving ability" (p. 177). According to *Corporate America* (Siegel and Sorensen 1994, 4) "universities are doing a less than adequate job" of preparing students for the accounting profession. As part of the CPA Vision Project (1998, 8), CPA focus groups found that "traditional education and training for the designation lack the breath of knowledge and skills required in the workplace."

These recommendations challenge accounting programs to prepare students to become independent learners, able to learn on their own. In addition, the recommendations call on accounting students to take a more active role in the learning process rather than being passive recipients of information. In fact, the Bedford Committee (AAA 1986, 169) emphasizes directing students' attention to the learning process and helping them learn how to learn "as the primary classroom objective." In *Objectives of Education for Accountants: Position Statement Number One* (AECC 1990), the AECC maintains that, due to a rapidly changing profession, accounting education programs no longer can prepare graduates to *be* accountants when they enter the profession. Instead, accounting education programs must prepare graduates to *become* professional accountants through a lifetime of experience, growth, and learning. Therefore, the AECC recommends that accounting education programs shift from a transfer of knowledge approach to a process-oriented approach placing new emphasis on learning how to learn (1990, 310).

The AECC (1990) and Francis, Mulder, and Stark (1995) submit that learning to learn can be conceptualized as a process of acquiring, understanding, and using a variety of learning strategies to help one learn more effectively. They propose that using these strategies improve one's ability to attain and apply knowledge, a process that enhances a lifelong desire to learn. Researchers from the field of education posit that learning strategies are an important aspect of the learning process and learning how to learn (Bransford 1979; Brown, Bransford, Ferrara, and Campione 1983; Jenkins 1979; Weinstein and Mayer 1986) and that learning strategies can be taught (Dansereau 1985; McKeachie, Pintrich, and Lin 1985; Weinstein and Underwood 1985). The research to date has investigated the impact of using learning strategies as instructional approaches to presenting new material. Is there potential benefit from introducing such tools and techniques as strategies that students can use from an individual perspective? The purposes of this paper are to describe learning strategies and suggest how accounting instructors can integrate several of these learning strategies in accounting courses. Also, encouraging students to use learning strategies may enable the students to take a more active role in their learning.

BACKGROUND

The Learning Process

Learning is a process that occurs when experience causes an essentially permanent change in an individual's knowledge or behavior (Biehler and Snowman 1993, G-4; Woolfolk 1998, 204). Views of learning have evolved from a *behavioral approach*, where learners passively receive knowledge, to a *cognitive approach*, where learners actively construct their knowledge. In the past two decades, the cognitive theories of learning have had the greatest impact on educational thought (Hergenhahn and Olson 1993, 464; Ornstein 1990, 33).

Cognitive theories of learning emphasize the individual learner in the learning process and are based on the cognitions (thoughts and mental processes) of learners as they encode, store, and retrieve information. The term metacognition refers to the individual learner's awareness and knowledge about these mental processes and how best to use these processes to achieve a learning goal. Learners can use many cognitive processes enabling them to take active control of their own learning (Weinstein and Meyer 1991). Instead of viewing the outcome of learning as depending mainly on what the teacher presents (through teaching strategies), the outcome depends jointly on what information the instructor presents and how the learner processes that information (through learning strategies) (Kardash and Amlund 1991, 117; Weinstein and Mayer 1986, 316).

Learners process information by paying attention to the information, recognizing it, transforming it, storing it in memory, and later retrieving it from memory (see Appendix for a discussion of information processing). The learners then store mental structures of organized knowledge, known as cognitive structures, in their memory (McKeachie et al. 1990). Individual learners construct different cognitive structures based on the experiences and expectations that influence their learning process and, therefore, learn differently. The cognitive approach to learning places greater emphasis on individual learners. Learners are no longer passive recipients of information, but potentially active participants in learning (Svinicki 1991, 27; 1994, 335). Learning strategies help learners process information and thus, become active participants in their learning.

Models of Learning

Learning models depict the various factors that influence the learning process. Several learning models from the educational literature have provided a basis for accounting education researchers in developing models for accounting education: Bloom's (1956) taxonomy of educational objectives (Needles and Anderson 1991); Perry's (1970) scheme of intellectual and moral development (Smith and Smith 1991); and Kolb's (1984) Experiential Learning Model (Baker, Simon, and Bazeli 1986, 1987; Baldwin and Reckers 1984; Jensen 1996). Of particular interest, Jenkins (1979) and Bransford (1979) present a general model of learning, grounded in the developmental and cognitive psychology literature, that offers a broad perspective of the learning process. Rather than a sequential flow chart of the learning cycle or a hierarchy of learning objectives, their model, represented as a tetrahedron, considers the interactive relationship of factors affecting the learning process.

Figure 1 highlights the tetrahedral model's interactive relationship among four fundamental factors affecting the learning process: two concern the learner and two concern the task environment. The first factor, *characteristics of the learner*, refers to the preexisting knowledge and capacities that the learner brings to the learning situation. These include individual differences in intelligence, personality, motivation, attitude, and learning style. The second factor, *learner activities (strategies)*, includes activities the learner engages in, spontaneously, or under another's guidance, when presented with a task. These two factors underscore that learning is an individual process. The third factor involves the *nature of the learning materials*. Bransford (1979) refers to the nature of materials as its modality (i.e., visual, linguistic). Nelson (1996) expands this factor to include both the materials to be learned and the manner in which the materials are presented. The fourth factor, *nature of the criterion or criterial tasks*, refers to the manner in which the learner is expected to demonstrate competence. As a general framework for conceptualizing the complexities of the learning process, Nelson (1996, 228) suggests that the tetrahedral model offers important insights into accounting education.

The tetrahedral model illustrates a relationship among the major factors that should be taken into account when considering any aspect of learning (Brown et al. 1983; Brown, Campione, and Day 1980). It suggests that for individuals to become independent learners, they need to know something about their own characteristics, their available learning strategies, the demand characteristics of various learning tasks, and the inherent structure of materials. They must adapt their learning strategies to the changing and competing demands of all these factors to be flexible and effective learners. In other words, they must "learn how to learn" (Brown et al. 1983, 106). Brown et al. stress the significance of learning strategies in students learning how to learn. Thus, identifying and describing the various learning strategies that students may use are important.

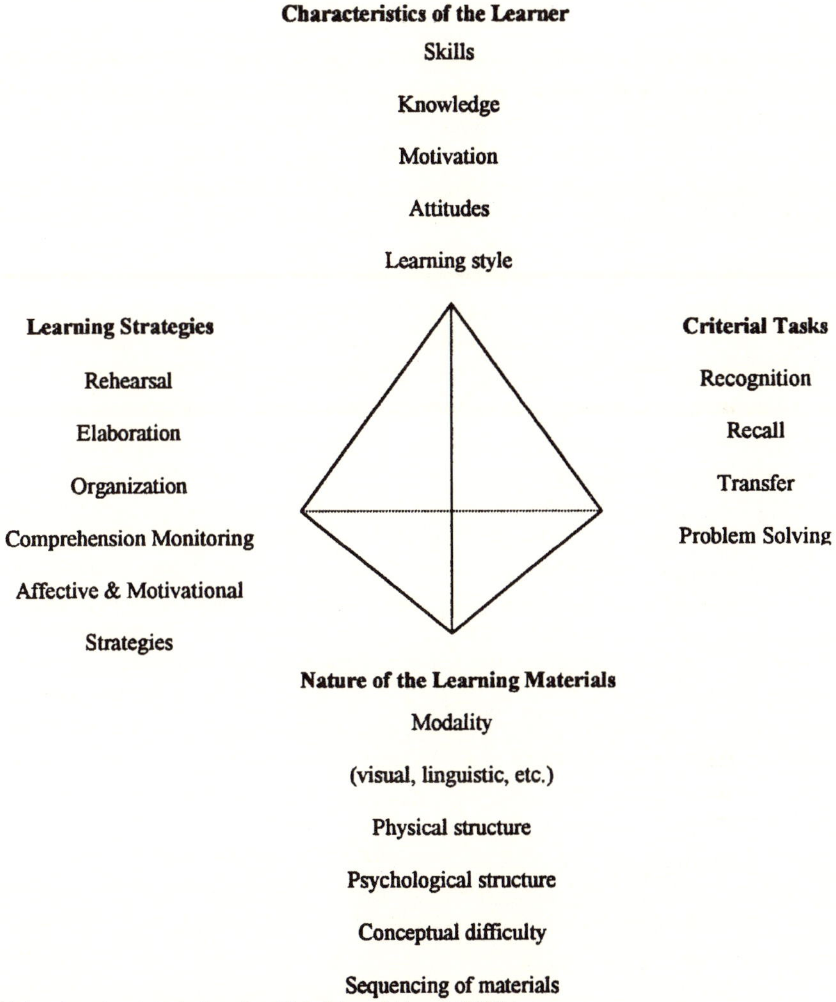

Characteristics of the Learner

Skills

Knowledge

Motivation

Attitudes

Learning style

Learning Strategies   **Criterial Tasks**

Rehearsal   Recognition

Elaboration   Recall

Organization   Transfer

Comprehension Monitoring   Problem Solving

Affective & Motivational

Strategies

Nature of the Learning Materials

Modality

(visual, linguistic, etc.)

Physical structure

Psychological structure

Conceptual difficulty

Sequencing of materials

Note: Learning strategies based on Weinstein and Mayer's (1986) taxonomy.
Source: Adapted from Jenkins (1979) and Bransford (1979).

Figure 1.   Tetrahedral Model of Learning

Learning Strategies

Researchers cannot agree on a precise definition of learning strategies (Dansereau 1985, 209; McKeachie et al. 1985, 154; Paris, Lipson, and Wixson 1983, 294; Tobias 1982, 5), however, they do agree on the importance and utility of learning strategies. A basic definition of a learning strategy is a plan or a sequence of mental activities designed to help reach a learning goal (Klauer 1988, 11; Weinstein and Meyer 1991, 17). It is not a simple action, but a complex ordered chain of actions, consisting of techniques, skills, or methods. Gagné (1974, 4) defines cognitive learning strategies as "skills of self-management that the learner acquires, presumably over a period of years, to govern his own processes of attending, learning, and thinking. By acquiring and refining such strategies, the student becomes an increasingly skillful independent learner and independent thinker."

Weinstein and Mayer (1986) present a broad definition encompassing many cognitive processes. They define learning strategies as "behaviors and thoughts that a learner engages in during learning and that are intended to influence the learner's encoding process" (1986, 315). Weinstein and Mayer (1986, 316) posit that instruction in learning strategies can affect learner characteristics by making specific strategies and methods available to the learner. The use of particular learning strategies can affect the encoding process, which in turn affects learning outcomes and performance.

Weinstein and Mayer (1986) created a taxonomy to describe major categories of learning strategies along with methods designed to influence aspects of the encoding process. Their taxonomy includes the categories of *rehearsal (repetition), elaboration, organization, comprehension monitoring*, and *affective and motivational strategies*.

Effective *rehearsal learning strategies* are appropriate when the instructional task requires simple recall or identification of important facts (Weinstein and Meyer 1991, 20). This category of learning strategy is found most often in introductory courses. Here the acquisition of basic knowledge is the first step in creating a more integrated knowledge base of an area. When the learning task involves knowledge and skills that go beyond a superficial level, the learner can use more complex rehearsal strategies. Examples include highlighting class notes or copying key ideas from a required reading. In short, "Repetition strategies are most appropriate in the early stages of building a base of knowledge in an area" (Weinstein and Meyer 1991, 20).

Based on the information processing model (see Appendix), moving information into long-term memory requires building a bridge between what one already knows and what one is trying to learn. *Elaboration learning strategies* allow learners to use existing knowledge or experiences to make what they are trying to learn more meaningful and memorable. Examples of elaboration learning strategies for basic tasks, such as learning lists, include the mnemonic keyword method or forming a mental image. If the learner is more active in building connections (e.g., finding similarities or differences) between existing knowledge and new information, the new knowledge becomes more meaningful and memorable. Elaboration strategies that require more active processing on the learner's part include paraphrasing, summarizing, generative note-taking, and creating analogies. In short, "The processes of comparing and contrasting help us to build connections to the new material, so that it is more easily moved into long-term memory" (Weinstein and Meyer 1991, 20).

Organization learning strategies include grouping or clustering items into categories, outlining a passage, or creating a hierarchy or concept map. These learning strategies require the translation or transformation of information into another form creating a structure for this new information. By organizing several separate pieces of information into a few, the new structure enables complex tasks to become more manageable and more meaningful, facilitating the move into long-term memory. The structure then serves as a guide back to the information when needed. The facilitating effect of both elaboration and organization learning strategies is attributed to both the process involved in applying the strategy and the outcome of the strategy. When learners use elaboration or organization learning strategies, "the mental activities that they engage in help them learn and remember the information" (Weinstein and Meyer 1991, 21).

Metacognition refers to both learners' knowledge about their own cognitive processes and their ability to control these processes by planning, monitoring, and evaluating them as a function of learning outcomes. The use of metacognitive strategies is often operationalized as comprehension monitoring. *Comprehension monitoring strategies* require the learner to establish learning goals with respect to an instructional task, to determine if these goals are being met, and, if necessary, to modify the strategies being used to meet these goals (Weinstein and Mayer 1986, 323). These strategies include the use of self-testing and questioning to help the learner in understanding the material and integrating it with existing knowledge.

Affective and motivational strategies are designed to help the learner focus attention, maintain concentration, manage performance anxiety (e.g., test anxiety), establish and maintain motivation, and manage time effectively (Weinstein and Mayer 1986, 324). Examples of affective strategies include reducing external distractions by studying in a quiet place, using thought stopping to prevent negative thoughts that disrupt performance, and using positive self talk (317).

The preceding description of learning strategies presents the wide variety of learning strategies available to students. Students' knowledge of learning strategies may include knowing that strategies exist and knowing how to use them, but often students do not know how to use strategies in the most effective manner.

In general, Paris et al. (1983) categorize these types of knowledge as: *declarative, procedural*, and *conditional knowledge*. *Declarative knowledge* is "knowing that" something is the case. In the case of learning strategies, it includes knowing about task characteristics, personal abilities, and various learning strategies. For example, "knowing that" summarizing is a type of elaboration learning strategy is declarative knowledge. *Procedural knowledge* involves "knowing how" to do things—it must be demonstrated. In the case of learning strategies, the learner must be able to summarize a passage. However, declarative and procedural knowledge alone are not sufficient. They only emphasize the knowledge and skills required for performance and do not address the conditions under which one might wish to select or execute actions. *Conditional knowledge* captures the dimension of "knowing when and why" to apply one's declarative and procedural knowledge.

Depending on the learners' characteristics, the nature of the learning task, and the nature of the materials, certain learning strategies may be more effective than others in various circumstances. If students have conditional knowledge of why some particular strategy works, they will be more likely to use it in an appropriate situation (Paris et al. 1983, 304). Individuals who can develop effective methods (i.e., using learning strategies) for adapting and regulating their information processing in handling various situations are better prepared to be lifelong learners (McKeachie et al. 1985, 153).

INTEGRATING LEARNING STRATEGIES

Although Weinstein and Mayer's (1986) taxonomy includes five categories of learning strategies, this paper presents suggestions for the integration of three: elaboration, organization, and comprehension monitoring. I exclude rehearsal strategies because, as indicated earlier, they are most appropriate in the early stages of building a knowledge base and have been proven in the literature to be the least effective of all strategies. The current paper focuses on those learning strategies which help build connections enabling students to move information into long-term memory. I also exclude affective and motivational strategies.

Elaboration

Elaboration strategies help learners to elaborate on new material to relate it to other information in the course material and/or information that the learners already posses (Pintrich et al. 1991; Shuell 1988). Paraphrasing, summarizing, and explaining to others are examples of elaboration strategies that can help learners move information into long-term memory by connecting and integrating new information with existing knowledge. The more active the learner in building these connections, the more memorable the new knowledge becomes (Weinstein and Meyer 1991, 20). Research by psychologists suggests that the effectiveness of elaboration appears to depend on a number of factors. For example, elaborations that clarify the significance of relationships appear to enhance learning more than other types of elaboration (Bransford et al. 1982; Stein et al. 1982), especially if learners generate elaborations more congruent with their existing knowledge base (Pressley et al. 1987).

Schadewald and Limberg (1990) were among the first to extend elaboration research into an accounting setting. They examined whether or not actively involving students in the process of determining the rationale for a rule, through self-generated elaborations, would enhance the students' ability to access and recall that rule later. Evidence suggests that self-generated elaborations tend to be more congruent with a learner's existing knowledge and, thus, are more extensive and better integrated elaborations. Because instructor-provided elaborations come from the instructor's knowledge base, their effectiveness depends largely on the congruence between the existing knowledge of the instructor and the learner (Pressley et al. 1987). The instructor, however, can provide students with several questions to guide the students in generating elaborations. Schadewald and Limberg (1990, 37) found that self-generated elaboration provided greater learning gains as students were better able to recall tax rules when they attempted to determine the rationale for tax rules on their own.

Hermanson (1994) extended Schadewald and Limberg's study to investigate whether or not accounting students' ability had an impact on the benefit associated with using self-generated elaboration. The results indicate that both high and low ability accounting students benefitted from the use of self-generated elaborations (315). Hermanson also found a positive association between elaboration reasonableness (accuracy of students' self-generated elaborations) and recall accuracy. If students elaborate new information by making incorrect connections or developing misguided explanations, these misconceptions will be remembered too. Thus, elaboration reasonableness is crucial to linking new information to existing knowledge, resulting in improved recall accuracy (Pressley et al. 1987; Stein et al. 1982; Woolfolk, 1998).

Instructors can encourage students to use elaboration learning strategies throughout the semester. Students may find them particularly helpful when introduced to a new chapter or topic. Johnson et al. (1991, 21) suggest providing students three to five minutes at the end of class to review and elaborate their notes and to ask clarifying questions of one another and/or the instructor. One-minute papers (Almer, Jones, and Moeckel 1998; Cottell and Harwood 1998; Stocks, Stoddard, and Waters 1992) are a variation of their suggestion. Instructors can ask the students to write a summary of the major points or issues presented in class and a question on a point that is unclear to the student. Instead of using class time, instructors can encourage students to ask themselves the following four questions (or variations thereof) within two hours after class:

a. What was the main point of the lesson?
b. What in the lesson did I find most interesting?
c. What is one probable test question that will come out of the lesson?
d. What one question do I most want to ask of my instructor? (Johnson et al. 1991, 21)

By reviewing the material within two hours, students reduce the effect of rapid forgetting. Using these examples instructors can encourage students to build the connections important to processing new information.

Since elaboration reasonableness (accuracy) is crucial to linking new information to existing knowledge, it is important for the instructor to review the students' elaborations. These examples enable the instructor to see if the students understand the material and to provide feedback to the students. If providing individual feedback is not possible, the instructor can clarify any misconceptions at the start of the next class session. Choo and Tan (1995) found that students using self-generated elaboration followed by instructor-assisted elaboration as a form of feedback produced the highest level of information processing.

As another suggestion, instructors can encourage students to work on problems in study pairs (or groups) where they take turns summarizing and explaining the concepts and issues to one another. This procedure compels the students to be more active and involved in their learning. Initially, it might be necessary to assign required problems to study pairs allowing students to form their own pairs later on. A good way to understand a concept better is to explain it to someone else. This allows for any discrepancies between the students' interpretations to be discussed and ultimately resolved, potentially through the instructor's clarification.

Instructors also should include essay questions on tests and examinations. The presence of essay items encourages the students to get an overall view of the material covered and to make connections among the many specific facts.

Organization

Organization strategies, including outlining chapters, creating concept maps and organizing the material, can help the learner identify main ideas and important supporting details from the material to be learned (Weinstein and Mayer 1986). Organizing is an active, effortful endeavor, requiring the learner to transform the main ideas and supporting details into another form or structure that is easier to understand. The purpose is to create added meaning for the information so that it will move into long-term memory more effectively. Remembering information that is structured is easier than remembering isolated pieces of information without structure (Weinstein and Meyer 1991, 20).

Organization includes techniques for grouping lists of items or for recognizing and recalling the structure of the information. Instructors can encourage students to create note cards capturing the key ideas, broad concepts, and relationships in the material to be learned. As instructors provide more detail about the new material, students can add new cards that build a structure to the information. Instructors should review the note cards occasionally and provide the students feedback. The note cards provide extended external storage that allows students to go back and review.

Some students may find visual strategies helpful. Evidence suggests that creating graphic organizers such as charts, diagrams, or concept maps are more effective than outlining or note taking (Robinson and Kiewra 1995). Visual maps show the connections and the underlying concepts that make or support the proposition.

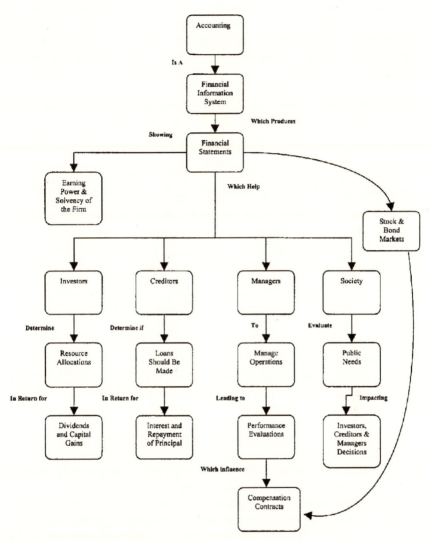

Source: Leauby and Brazina (1988).

Figure 2. Example of Concept Map Used in
Introductory Accounting Course

Thus, a concept map is a schematic representation of meaningful relationships
between concepts that makes the information easier for students to understand and
encourages further independent exploration.

Leauby and Brazina (1998) describe how instructors and students can use concept mapping in accounting courses to develop students' capacity to learn independently. When assigning or discussing a new topic, use the concept map as a road map or overview of the topic. Upon completing discussion of the topic, the map provides a schematic summary of what students hopefully learned. Figure 2 presents an example of a concept map used in introductory accounting.

Another approach to using concept maps is after the students have read the new material (e.g., a journal article, a chapter in the text), ask them to select and describe the main concepts. This can be done individually, in small groups, or as a class exercise. Instructors can start the process by asking students to set up the hierarchy of the concept map by having the students rank the concepts, starting with the most inclusive or comprehensive concept. This activity may result in different rankings based on the students' perceptions. After agreeing on the rankings, students can begin to build the concept map using appropriate linking words to make connections between the concepts (Leauby and Brazina 1998, 130). After the first creation, a period of reflection allows students to exchange ideas about the validity of certain relationships. Also, this period allows for the discovery of missing linkages. The greatest benefits of concept mapping are gained when students get actively involved in their own learning process and construct and modify their own concept maps (126).

Organization strategies, including note cards and concept maps, help students transform the main ideas and supporting details into a structure that is easier for them to understand. Instructors should encourage students to use strategies that help identify these main ideas and supporting details from the material to be learned.

Comprehension Monitoring

Comprehension monitoring strategies include metacognitive self-regulation activities. Metacognition refers to (1) the awareness of and knowledge about cognition and (2) the control and regulation of cognition (Brown et al. 1983, 107). Planning, monitoring, and regulating processes make up the metacognitive self-regulation activities. Planning activities occur before undertaking a task and include goal setting and task analysis. These help the learner activate relevant aspects of existing knowledge that facilitate organizing and comprehending material. Monitoring activities occur during learning and include tracking one's attention, self-testing, and questioning while reading or listening to lectures. These activities can help the learner in understanding the material and integrating it with existing knowledge. Self-regulating activities check the outcomes and include continuous adjustment and fine-tuning one's cognitive activities. These activities are assumed to enhance performance by helping learners in checking and correcting their behavior as they proceed on a task (McKeachie et al. 1990, 33).

Preview	Introduce yourself to each new chapter by surveying the major topics and sections. Read the overview, objectives, section headings and subheadings, summary, and perhaps the initial sentences of the major sections. These procedures will help activate schemas so you can interpret and remember the text that follows.
Question	For each major section, write questions that are related to your reading purposes. One way is to turn the headings and subheadings into questions.
Read	The questions you have formulated can be answered through reading. Pay attention to the main ideas, supporting details, and other data in keeping with your purposes. You may have to adjust your reading speed to suit the difficulty of the material and your purpose in reading.
Reflect	While you are reading, try to think of examples or create images of the material. Elaborate and try to make connections between what you are reading and what you already know.
Recite	After reading each section, sit back and think about your initial purposes and questions. Can you answer the questions without looking at the book? In doing this, you give your mind a second chance to connect what you have read with what you already know. If your mind is blank after reading a section, it may have been too difficult to read comfortably, or you may have been daydreaming. Reciting helps you to monitor your understanding and tells you when to reread before moving on to the next section. Reciting should take place after each headed section.
Review	Effective review incorporates new material more thoroughly into your long-term memory. As study progresses, review should be cumulative, including the sections and chapters you read previously. Rereading is one form of review, but trying to answer key questions without referring to the book is the best way. Wrong answers can direct you to areas that need more study, especially before an exam.

Figure 3. Learning from Reading–PQ4R

A learning technique that incorporates planning, monitoring, and regulating activities is the PQ4R method (Thomas and Robinson 1972, cited in Woolfolk 1998). Typical accounting textbooks are very different from textbooks of other disciplines. Adelberg and Razek (1984) found that accounting textbooks are difficult to read and they described the concept explanations as difficult to comprehend. The PQ4R method is a useful technique to enhance students' learning from reading. Figure 3 identifies and describes the steps included in the PQ4R method.

Instructors should take time at the beginning of a course to go over the method. In previewing a new chapter, instructors can point out the learning objectives and section headings. When reading the chapter, instructors should encourage students to make connections between the new material and their existing knowledge. After reading a section, instructors can ask students to think about what they have read to monitor their own understanding. Instructors should stress the importance of incorporating these steps in the reading of accounting material.

Following these steps students can become more aware of the organization of the material. The headings provide clues to how the information is organized. These steps also require students to study the material in sections instead of trying to learn all the information at once. Asking and answering one's own questions about the material encourages students to process the information more deeply and with greater elaboration (Woolfolk 1998, 313). Students also can use the one-minute papers as a comprehension monitoring strategy to check their understanding of the new material. The steps in the PQ4R and the one-minute papers provide students an opportunity to build connections between the new material and their existing knowledge better.

CONCLUSIONS

The suggestions presented in this paper are in response to criticisms that accounting education is not preparing students for a lifetime of learning. The AECC states that students should be taught skills and strategies that help them learn more effectively and how to use these effective learning strategies to continue to learn throughout their lifetime. To become independent learners, students need to know something about their own characteristics, their available learning strategies, the demand characteristics of various learning tasks, and the inherent structure of materials. They must adapt their learning strategies to the changing and competing demands of all these factors to be flexible and effective learners.

Providing students with the knowledge of various learning strategies and how to use them improves the students' opportunity to learn independently. This paper presents examples of several learning strategies that could be integrated into accounting courses. However, integrating all these learning strategies in every accounting course would be impossible. Instructors need to consider their course content when evaluating which learning strategy or strategies to integrate.

Based on my initial efforts and observation of others attempting to integrate learning strategies, instructors should integrate learning strategies from the beginning of the course to gain the greatest potential impact on their use. Beginning early in the semester allows for a more in depth discussion of various learning strategies, their use, and their appropriateness for given learning tasks. It also provides more opportunity for the students to practice and implement their use. Through accounting related examples, instructors can model the use of learning strategies such that students see how learning strategy use can enhance the learning of new material.

Several potential problems exist with integrating learning strategy use:

1. How can learning strategy use be measured?
2. Can a student's learning activities be changed or modified at this level?
3. What is the motivation for students to learn and implement these strategies?
4. If learning strategy use enhances learning, how are learning outcomes measured?

This paper proposes that there is a benefit from introducing these tools and techniques as strategies that the students can use from an individual perspective to enhance their learning. Further research is needed with respect to the implementation of the suggestions made here.

APPENDIX

Information Processing System

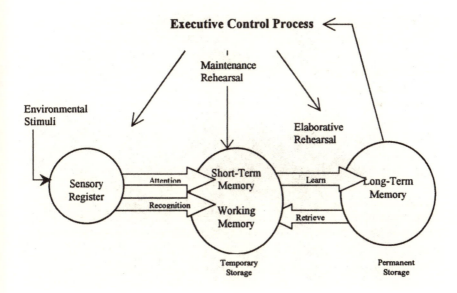

Source: Adapted from Biehler and Snowman (1993) and Woolfolk (1998).

According to the information processing view, learning results from an interaction between an environmental stimulus (information that is to be learned) and a learner (the one who processes, or transforms, the information). The information processing model identifies three memory stores (sensory, short-term, and long-term), each varying as to the amount of information they can hold and for how long. To move information between the memory stores, the learner needs to generate connections between the information to be learned and the knowledge already organized in his/her memory file. These connections may include similarities, differences, structure, or hierarchies in the new material that are similar or different from a knowledge base already understod by the learner. The executive control processes govern both the manner in which information is encoded and its flow through the information processing system. Elaborative rehearsal is a general term including various organizational and meaning-enhancing encoding techniques. The control processes are important to the information processing system. First, they determine the quantity and quality of information that the learner stores and retrieves from memory. Second, it is the learner who decides whether, when, and how to employ them (Biehler and Snowman 1993, 382; Weinstein and Meyer 1991, 16).

ACKNOWLEDGMENT

I would like to thank the editor, Bill N. Schwartz, for his valuable comments and suggestions and two anonymous reviewers for their constructive comments.

NOTE

1. The Accounting Education Change Commission, formed in 1989 to help bring accounting education into the twenty-first century, completed its mission in the Fall of 1996.

REFERENCES

Accounting Education Change Commission (AECC). 1990. Objectives of Education for Accountants: Position Statement No. 1. *Issues in Accounting Education* (Fall): 307-312.

Accounting Education Change Commission (AECC). 1992. The First Course in Accounting: Position Statement No. 2. *Issues in Accounting Education* (Fall): 249-251.

Adelberg, A., and J. Razek. 1984. The cloze procedure: A methodology for understanding the understandability of accounting textbooks. *The Accounting Review* (January): 109-122.

Almer, E.D., K. Jones, and C.L. Moeckel. 1998. The impact of one-minute papers on learning in an introductory accounting course. *Issues in Accounting Education* (August): 485-497.

American Accounting Association (AAA) Committee on the Future Structure, Content, and Scope of Accounting Education (Bedford Committee). 1986. Future accounting education: Preparing for the expanding profession. *Issues in Accounting Education* (Spring): 168-195.

Arthur Andersen & Co., Arthur Young, Coopers & Lybrand, Deloitte Haskins & Sells, Ernst & Whinney, Peat Marwick Main & Co., Price Waterhouse, and Touche Ross. 1989. *Perspectives on Educational Capabilities for Success in the Accounting Profession* (Big-Eight White Paper). New York: Authors.

Baker, R.E., J.R. Simon, and F.P. Bazeli. 1986. An assessment of the learning style preferences of accounting majors. *Issues in Accounting Education* (Spring): 1-12.

Baker, R.E., J.R. Simon, and F.P. Bazeli. 1987. Selecting instructional design for introductory accounting based on the experiential learning model. *Journal of Accounting Education* 5: 207-226.

Baldwin, B.A., and P.M.J. Reckers. 1984. Exploring the role of learning style research in accounting education policy. *Journal of Accounting Education* 2: 63-76.

Biehler, R.F., and J. Snowman. 1993. *Psychology Applied to Teaching*, 7th ed. Boston: Houghton Mifflin Company.

Bloom, B.S., ed. 1956. *Taxonomy of Educational Objectives*. New York: David McKay Company, Inc.

Bransford, J.D. 1979. *Human Cognition: Learning, Understanding and Remembering*. Belmont, CA: Wadsworth Publishing Company.

Bransford, J.D., B.S. Stein, N.J. Vye, J.J. Franks, P.M. Auble, K.J. Mezynski, and G.A. Perfetto. 1982. Differences in approaches to learning: An overview. *Journal of Experimental Psychology: General* 111: 390-398.

Brown, A.L., J.D. Bransford, R.A. Ferrara, and J.C. Campione. 1983. Learning, remembering and understanding. In *Handbook of Child Psychology*, Vol. 3, edited by P.H. Mussen. New York: John Wiley & Sons.

Brown, A.L., J.C. Campione, and J.D. Day. 1980. *Learning to Learn: On Training Students to Learn From Text* (Technical Report No. 189). Bethesda, MD: National Institute of Child Health and Human Development. (ERIC Document Reproduction Service No. ED 203 297) (*Educational Researcher, 10*(2), 14-21, 1991).

Choo, F., and K. Tan. 1995. Effect of cognitive elaboration on accounting students' acquisition of auditing expertise. *Issues in Accounting Education* (Spring): 27-45.

Cottell, P.G., Jr., and E.M. Harwood. 1998. Using classroom assessment techniques to improve student learning in accounting courses. *Issues in Accounting Education* (August): 551-564.

CPA Vision 2011 and Beyond: Focus on the Horizon. 1998. *Journal of Accountancy* (December): 25-73.

Dansereau, D.F. 1985. Learning strategy research. In *Thinking and Learning Skills: Relating Instruction to Research*, Vol. 1, edited by J.W. Segal, S.F. Chipman, and R. Glaser. Hillsdale, NJ: Lawrence Erlbaum.

Francis, M.C., T.C. Mulder, and J.S. Stark. 1995. *Intentional Learning: A Process for Learning to Learn in the Accounting Curriculum*. Accounting Education Series, Volume No. 12. Sarasota, FL: American Accounting Association.

Gagné, R. 1974. Educational technology and the learning process. *Educational Researcher* 3: 3-8.

Hergenhahn, B.R., and M. Olson. 1993. *An introduction to theories of learning*, 4th ed. Englewood Cliffs, NJ: Prentice-Hall.

Hermanson, D.R. 1994. The effect of self-generated elaboration on students' recall of tax and accounting material: Further evidence. *Issues in Accounting Education* (Fall): 301-318.

Jenkins, J.J. 1979. Four points to remember: A tetrahedral model of memory experiments. In *Levels of Processing in Human Memory*, edited by L.S. Cermak and F.I.M. Craik. Hillsdale, NJ: Lawrence Erlbaum.

Jensen, P.H. 1996. The application of Kolb's Experiential Learning Theory in a first semester college accounting course (Ph.D. dissertation, University of Memphis, 1995). *Dissertation Abstracts International*, 56, 3201A.

Johnson, G.R., J.A. Eison, R. Abbott, G.T. Meiss, J.A. Morgan, T.L. Pasternack, E. Zaremba, and W.J. McKeachie. 1991. *Teaching Tips for Users of the Motivated Strategies for Learning Questionnaire (MSLQ)* (Report No. NCRIPTAL-91-P-005). Ann Arbor, MI: National Center for Research to Improve Postsecondary Teaching and Learning. (ERIC Document Reproduction Service No. ED 338 123).

Kardash, C.M., and J.T. Amlund. 1991. Self-reported learning strategies and learning from expository text. *Contemporary Educational Pyschology* 16: 117-138.

Klauer, K.J. 1988. *Teaching for Learning-to-learn: A Critical Appraisal with Some Proposals*. Paper presented at the Annual Meeting of the American Educational Research Association. New Orleans, LA.

Kolb, D. 1984. *Experiential Learning*. Englewood Cliffs, NJ: Prentice-Hall.

Leauby, B.A., and P. Brazina. 1998. Concept mapping: Potential uses in accounting education. *Journal of Accounting Education* 16: 123-138.

McKeachie, W.J., P.R. Pintrich, and Y. Lin. 1985. Teaching learning strategies. *Educational Psychologist* 20: 153-160.

McKeachie, W.J., P.R. Pintrich, Y. Lin, D.A.F. Smith, and R. Sharma. 1990. *Teaching and Learning in the College Classroom: A Review of the Research Literature*, 2nd ed. Ann Arbor: University of Michigan Press.

Needles, B.E., Jr., and H.R. Anderson. 1991. A comprehensive model for accounting education. In *Models of Accounting Education*, edited by G.L. Sundem and C.T. Norgaard. Torrance, CA: AECC.

Nelson, I.T. 1996. A tetrahedral model of accounting education: A framework for research. *Journal of Accounting Education* 14: 227-236.

Ornstein, A.C. 1990. *Strategies for effective teaching*. New York: Harper & Row.

Paris, S.G., M.Y. Lipson, and K.K. Wixson. 1983. Becoming a strategic reader. *Contemporary Educational Psychology* 8: 293-316.

Perry, W.G., Jr. 1970. *Forms of Intellectual and Ethical Development in the College Years: A Scheme*. New York: Holt, Rinehart & Winston.

Pintrich, P.R., D.A.F. Smith, T. Garcia, and W.J. McKeachie. 1991. *A Manual for the Use of the Motivated Strategies for Learning Questionnaire (MSLQ)*. Ann Arbor: University of Michigan, National Center for Research to Improve Postsecondary Teaching and Learning.

Pressley, M., M.A. McDaniel, J.E. Turnure, E. Wood, and M. Ahmad. 1987. Generation and precision of elaboration: Effects on intentional and incidental learning. *Journal of Experimental Psychology: Learning, Memory, and Cognition* 13: 291-300.

Robinson, D.H., and K.A. Kiewra. 1995. Visual arguments: Graphic outlines are superior to outlines in improving learning from text. *Journal of Educational Psychology* 87: 455-467.

Schadewald, M., and S. Limberg. 1990. Instructor-provided versus student-generated explanations of tax rules: Effect on recall. *Issues in Accounting Education* (Spring): 30-40.

Shuell, T.J. 1988. The role of the student in learning from instruction. *Contemporary Educational Psychology* 13: 276-295.

Siegel, G., and J.E. Sorensen. 1994. *What Corporate America Wants in Entry-level Accountants: Results of Research*. Montvale, NJ: Institute of Management Accountants.

Smith, G.S., and C.W. Smith. 1991. A working paper of a model of undergraduate accounting education for life-long learning. In *Models of Accounting Education*, edited by G.L. Sundem and C.T. Norgaard. Torrance, CA: AECC.

Stein, B.S., J.D. Bransford, J.J. Franks, R.A. Owings, N.J. Vye, and W. McGraw. 1982. Differences in the precision of self-generated elaborations. *Journal of Experimental Psychology: General* 111: 399-405.

Stocks, K.D., T.D. Stoddard, and M.L. Waters. 1992. Writing in the accounting curriculum: Guidelines for professors. *Issues in Accounting Education* (Fall): 193-204.

Svinicki, M.D. 1991. Practical implications of cognitive theories. In *College Teaching: From Theory to Practice*, edited by R. Menges and M.D. Svinicki. San Francisco: Jossey-Bass.

Svinicki, M.D. 1994. Research on college student learning and motivation: Will it affect college instruction? In *Student Motivation, Cognition, and Learning: Essays in Honor of Wilbert J. McKeachie*, edited by P.R. Pintrich, D.R. Brown, and C.E. Weinstein. Hillsdale, NJ: Lawrence Erlbaum.

Tobias, S. 1982. When do instructional methods make a difference? *Educational Researcher* 11(4): 4-10.

Weinstein, C.E., and R.E. Mayer. 1986. The teaching of learning strategies. In *The Handbook of Research on Teaching*, 3rd ed., edited by M. Wittrock. New York: Macmillan.

Weinstein, C.E., and D.K. Meyer. 1991. Cognitive learning strategies and college teaching. In *College Teaching from Theory to Practice*, edited by R.J. Menges and M.D. Svinicki. San Francisco: Jossey-Bass.

Weinstein, C.E., and V.L. Underwood. 1985. Learning strategies: The how of learning. In *Thinking and Learning Skills: Relating Instruction to Research*, Vol. 1, edited by J.W. Segal, S.F. Chipman, and R. Glaser. Hillsdale, NJ: Lawrence Erlbaum.

Woolfolk, A.E. 1998. *Educational Psychology*, 7th ed. Needham Heights, MA: Allyn and Bacon.

STUDENTS MAY BLOSSOM USING BLOOM'S TAXONOMY IN THE ACCOUNTING CURRICULUM

Julia K. Brazelton

ABSTRACT

Cognitive development should be education's ultimate goal. One methodology, Bloom's taxonomy in the cognitive domain (1956), accomplishes the educational goal of lifelong learning. Bloom's framework encourages students' thinking at the highest levels, fostering their ability to learn and their interest in learning. The expressed purpose of implementing Bloom's theory is to develop students' cognitive reasoning skills. This paper presents several methodologies designed to achieve higher-order cognate abilities. In addition to providing a framework for more effective teaching and learning, implementing Bloom's taxonomy makes the classroom environment more interesting. This paper describes an approach to incorporating oral and written communication in the accounting classroom and offers suggestions for incorporating Bloom's taxonomy into the accounting curriculum as well.

Advances in Accounting Education, Volume 2, pages 57-85.
ISBN: 0-7623-0515-0

INTRODUCTION

Liberating education only occurs when people develop their critical reasoning skills, including self-knowledge and self-awareness. This ability to think critically separates the autonomous, independent people, who are capable of making free choices, from the passive receivers of information. Liberating education consists of acts of cognition, not transferrals in information (Freire 1970, 67).

Currently, there is a movement, which began more than a decade ago, to incorporate critical thinking into the college classroom (AECC 1990; Nelson 1995). The accounting profession traditionally has been concerned with communication skills; however, critical thinking and communication skills are inherently interrelated. Fostering higher-order cognitive abilities should be incorporated into the ongoing dialogue about communication.

This paper links education literature about Bloom's taxonomy to accounting students' critical thinking and communication skills. In the past, the phrase "writing across the curriculum" was popular; now, "writing to learn" is the focus in educational spheres (Spear 1983; Cunningham 1996). The term, "writing to learn," is more accurate when referring to the purpose of writing. "The writing-to-learn movement is fundamentally about using words to acquire concepts" (Connolly and Vilardi 1989, 5). Findings of the American Association for the Advancement of the Humanities (1982) indicate that writing often is viewed as the end of the learning process rather than as a means of learning.

Oral communication skills also develop thinking skills; so, classroom discussions, in which students think aloud, offer opportunities for increased intellectual development. Learning is more than simply reiterating facts, clarifying meaning, and reinforcing concepts committed to memory. Learning involves making sense out of the material. Accounting educators should heed the experience of teachers in other disciplines and incorporate learning and thinking skills into their curricula. For example, Fulwiler (1987, 3) said

> If we are interested in helping...colleges do better what they are charged with doing--teaching people to reason systematically, logically, and critically--then we need to balance the curriculum as carefully with regard to (communication) activities as we currently do with (other) activities.

Developing communication skills in the accounting curriculum has been widely recognized as desirable. The AICPA (1988) and the Big 5 (Arthur Andersen et al., 1989) have emphasized the need to have competent speakers and writers in the accounting profession, while the AECC (1990) supports a commitment to lifelong learning. For more than three decades, accounting educators have recognized the importance of communication skills (Estes 1979; Henry and Razzouk 1988; May and Arevalo 1983). Many practitioners have indicated a willingness to sacrifice technical knowledge for improved writing skills (Novin and Pearson 1989). The American Accounting Association emphasizes the need for communication skills:

> Probably no other personal quality is more important to an accountant than having the ability to communicate well, both in writing and orally (AAA 1968).

Over the last decade, most accounting programs around the country have endorsed oral and written communication as part of their responsibility, and numerous authors have addressed the reasons for including written assignments and presentations in accounting programs (e.g., Laufer and Crosser 1990; Locke and Brazelton 1997). Communication and learning skills are not mutually exclusive; instead, they are complementary activities. Writing to learn improves written communication skills. In-depth class discussions require problem-solving, critical thinking, and evaluation activities, resulting in improved oral communication performance. These communication competencies are a direct result of improved learning.

Many faculty reluctantly have relented to the inclusion of student-generated communication in their classrooms (Phillips and Davis 1991). Some instructors claim that they are not equipped with the skills of English teachers and, therefore, are not qualified to teach writing and have been begrudgingly forced to make writing assignments. Few accounting professors have accepted the challenge of including writing and extensive discussion in their classes willingly. Fewer yet have endorsed communication exercises as an exciting opportunity to teach accounting differently. Accounting students need to write about and discuss topics to understand accounting fully. Further, if done properly, the communication process enables students' ability to learn throughout their lifetimes, but they must practice synthesizing and evaluating information and then communicating (both in oral and written form) their understanding to others in order to learn how to learn. Students must learn how to learn for themselves (Francis, Muldur, and Stark 1995).

Cognitive development follows hierarchical stages and indicates that designing curricula to promote critical thinking skills is possible (Spear 1983, 47). Existing frameworks present in educational literature easily could apply to accounting. There are problem-solving models with bases in psycholytical theories addressing the type of learning that is best for a particular group of students. Bloom (1956) suggests that students learn cumulatively at six levels of cognitive development, ranging from memorization of facts to evaluation of material. Cognitive development refers to the changes in thinking patterns that occur over time. Piaget (1972) suggests that instructors should adjust to the learner's stage of development in order to develop more complex levels of thinking in stages or schema to facilitate organization of the world in some way. The Myers-Briggs (1976) approach categorizes learning type indicators into sensing, intuitive, thinking, and feeling. Students base their preferences on their learning styles. For instance, an

Intuitive-Thinking student prefers a teacher who teaches using a compare and contrast method, while the Intuitive-Feeling participant prefers the "what if?" method. Perry (1970b), classifies learners according to their attitudes about learning, and Kohlberg (1970) espouses a model of learning that is based on an obedience and punishment orientation. The three stages of moral development are essential for an instructor to understand because, according to Kolberg's model, teaching plans should be organized with moral development as the guiding principle.

Despite differences in detail, these cognitive theorists concur about intellectual development. As thinking matures, it becomes "increasingly integrative, classificatory, and discriminating; progressively more independent of concrete referents; more abstract, hypothetical, and relativistic; and less egocentric and decreasingly subject to peer or authority pressure" (Spear 1983, 47-48). Bloom's taxonomy provides synergy between thinking-learning and curricular needs. Unlike other developmental models which are primarily chronological, Bloom's framework is less structured: it can apply equally to class periods as to semesters, or it can apply to elementary school children as well as to graduate students and faculty. Bloom (1956) provides a sequencing of assignments that assists students in their quest for higher understanding. Classroom communications, both verbal and written, promote cognitive growth. Thought processes used in one level are building blocks for the next level. The final cognitive plane, evaluation, builds on all previous stages. Conversion of information into some framework requires operation at higher levels. Students can learn more as active participants in written and verbal communication than as passive receptors of information. Questioning in the classroom and incorporating writing fully into the curriculum affect the learning process and are, indeed, methods of learning in themselves.

Communication skills evolve through training. "Eloquence is an acquired skill" (Civikly 1986, 8). Teaching is about interacting, both in the written and oral forms of communication. Teaching is not merely an act; instead, it is about a relationship in the classroom. The ability to think, talk, and write clearly are important. Students need to communicate clearly to think at higher, more creative levels and their construction of clear, logical arguments requires complete understanding of the material.

Bloom's taxonomy may foster the intellectual development of accounting students. The next section of this article examines the history of the taxonomy. Subsequent sections address the foundations of the taxonomy and illustrate methods available to accounting instructors to implement the highest levels of Bloom's taxonomy into the classroom.

HISTORY OF BLOOM'S TAXONOMY

Educators tout Bloom's taxonomy (1956) as one of the most influential educational monographs of the last fifty years. It is cited between one and two hundred times annually in *Social Sciences Citation Index* (Anderson and Sosniak 1994, vii). The purpose of Bloom's taxonomy is to distinguish between memorized knowledge and intellectual activities involving different abilities and skills. At the University of Chicago in the 1940s, undergraduate students took interdisciplinary courses and took comprehensive examinations on the material. Many examiners believed that the students spent too much time at the lowest level of thinking--knowledge of facts. The exams purported to specifically emphasize the higher mental processes. Bloom's idea to categorize the types of cognition required for the exam questions (to ascertain the portion allocated to higher levels of thinking) originated at the 1948 meeting of the American Psychological Association where it was proposed to a group of university examiners responsible for evaluation of their university programs. The structure was further developed at special meetings and subsequent conferences, directly incorporating the work of 30 people and subjected to the scrutiny of more than 1000 reviewers. Bloom's framework implicitly is dependent on the objectives of the educational process. Educators transformed the taxonomy into a basic reference tool worldwide, often enhanced for specialized purposes, and it allows educators to evaluate stages of student learning systematically. Bloom's taxonomy is responsible for differentiating learning. As Gebe (1990) noted about the taxonomy:

> Learning was learning, period; educators really didn't perceive it as a multileveled affair. The taxonomy was an organized way of thinking about the outcomes of instruction. In a way, the taxonomy marked the birth of instructional design as a subject or process that could be studied in a formal manner. Bloom's taxonomy describes the intended behaviors of students who have received instruction, not the actual behaviors. It's, therefore, a focus on input, not output.

If instructors use educational objectives when designing course curriculum, from a day's lecture to the course syllabus, they use Bloom's taxonomy, whether consciously or unconsciously.

Unlike stages of intellectual development utilized by most theorists (e.g., Perry 1970a; Kurfiss 1989), Bloom's taxonomy utilizes consecutive building on the structure. That is, a student may be in the first stage of intellectual development in which answers are right or wrong and lack ambiguity; nevertheless, the instructor can assist students in reaching higher orders of thinking such as synthesizing information. For instance, in an accounting principles course, an assignment that requires computation of financial ratios also could require interpretation and analysis of overarching issues.

Subject-specific taxonomies based on the Bloom framework were developed in the last 25 years in fields as diverse as art, history, math, music, and medicine. The taxonomy is not designed to impose teaching procedures (Anderson and Sosniak 1994). Its purpose is to afford many possibilities for instructional methodologies intended to accomplish the goals of each level. This approach does not involve rote-learning. It encourages students to learn and apply basic knowledge in increasingly more complex situations and circumstances.

FOUNDATIONS OF BLOOM'S TAXONOMY

The means through which the taxonomy provides intellectual growth appears in
Table 1. Knowledge is the basis for all cognitive processes; thus, it is needed for
all subsequent levels in Bloom's taxonomy. As knowledge expands, other levels
appear, with each higher-order cognitive skill requiring facts, then understanding,
etc. For instance, analysis is based on knowledge and comprehension and requires
the ability to apply precepts and analyze them; therefore, in the graphic, analysis
is a 4x5 rectangle passing through knowledge, comprehension, and application,
still requiring skills beyond the first three levels. The rectangle representing the
highest-order cognitive ability, evaluation, utilizes skills of all previous levels,
passing through all other rectangles in Figure 1.

The ability to classify questions into their respective categories in Bloom's paradigm is a necessary skill because critical thinking requires organization of
thoughts. Bloom offers a framework around which to organize classroom activities to accomplish educational goals the accounting profession advocates. The

Table 1.   Classification of Queries into Bloom's Taxonomy

Taxonomic Level	Key Words
Knowledge	List
	Name
	Define
	Describe
Comprehension	Explain
	Interpret
	Summarize
	Give examples
	Predict
	Translate
Application	Compute
	Solve
	Apply
	Modify
	Construct
Analysis	How does x apply?
	Why does x work?
	How does x relate to y?
	What distinctions can be made about x?
Synthesis	How do the data support x?
	How would you design a project that investigates x?
	What predictions can you make based on the data?
Evaluation	What judgments can you make about x?
	Compare and contrast x using criteria for y?

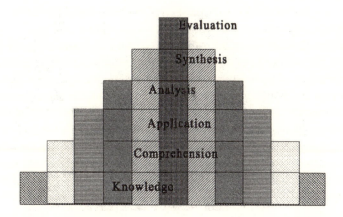

Figure 1. Bloom's Taxonomy

purposes of accounting education include teaching students how to learn and promoting lifelong learning. Table 1 presents a summary of key words for questions an instructor could ask at each taxonomic level.

Bloom's First Level: Knowledge

Bloom's Taxonomy of Educational Goals (1956) is divided into six distinct categories, each building on the previous item. The first level, ***knowledge***, includes behaviors and situations emphasizing remembering; thus, recognition or recall of material describes the learning process of this level. This most basic of concepts entails learning, storing, and recalling of data. Some alterations to the material should not affect the recollection of knowledge, particularly when the questions appear in a different form in a situation other than the original learning environment. This remembering skill is required in more advanced categories as well; however, the primary psychological process at this lowest level is remembering.

The three components of rudimentary knowledge are (1) knowledge of specifics, (2) knowledge of ways and means of dealing with specifics, and (3) knowledge of the universals and abstractions in a field. Generally, facts and terminology comprise the first sub-category, knowledge of specifics. Accounting, like other specialized fields of study, contains referents practicing accountants use. This "vocabulary" is a basic requirement to communicate in the field. Students, as well as practicing accountants, may be unable to consider economic phenomena unless they are able to utilize these terms and symbols. It is important at this point not to impose more terminology than the student can assimilate and retain.

In the second sub-category, the organizational methods for addressing the facts are more abstract than the first. This category is important for accounting students since it includes the conventions and rules of the field, including the interrelationships of the processes and methodologies involved and the ability to categorize. The classifications seem arbitrary to students in many fields, and accounting is no exception; nevertheless, practicing accountants find these categories invaluable and fundamental to their work. Instructors should expect students to know these categories at this level; however, application of these classifications to new situations exists at a higher taxonomic level.

The third sub-component of knowledge (universals and abstractions) encompasses the knowledge of principles, theories, and broad generalizations. Examples of items in this category of knowledge as applied to accounting education are: list four assets, define the matching principle, recall a particular SFAS, describe the basics of various forms of business ownership, and have an awareness of trends in capital gains laws over time.

Bloom's Second Level: Comprehension

Similarly, there are three components of Bloom's second level of learning: **comprehension.** These types of cognition include (1) translation, (2) interpretation, and (3) extrapolation. **Comprehension** is defined as "objectives, behaviors, or responses which represent an understanding of the literal message contained in a communication" (Bloom 1956, 89). The first component, translation, provides the means by which knowledge transitions to other types of learning and requires students to have the basic facts or prerequisite knowledge before engaging in more complex thinking about the idea. Abstract ideas must be converted into concrete terms the students already comprehend in order to "understand" at this level. Some students are able to perform translation while others are not. If a communication is not converted into briefer terms to facilitate thinking, students must rely on higher orders of cognition to accomplish this basic task. When instructors translate these concepts, students are to use recall. Asking students to translate an accounting principle into an example assures the student's ability to perform the task of comprehension. If the student can rephrase a problem presented in technical terms into lay terms, translation satisfies the requirements of comprehension.

The second component, interpretation, requires the ability to extend recognition between essential and more irrelevant points. That is, can the student identify the major points? Extrapolation, the third component, requires the learner to extend trends inherent in a communication to determine probable implications of the situation. If a student can select the correct answer to a test question when the term provided is different from that used in class or textual materials, the student can translate; however, if more than one new term is provided in a single problem, students able to translate may fail to select the appropriate term.

Bloom's Third Level: Application

The third level of thinking in Bloom's taxonomy is *application.* This level requires students to use the skills lower in the hierarchy effectively. Possession and application of knowledge are not synonymous. Training transfers to new concepts when the student learns methods of attacking problems. This skill is critical in dynamic fields such as accounting where constant presentation of new information requires application in a variety of diverse situations. Examples of application include the ability to apply accounting principles to new fact situations, to apply existing tax laws to a new set of facts, to interpret the effect of a current event, or to utilize an accounting principle in a concrete example.

Bloom's Fourth Level: Analysis

The fourth level of cognitive skills development in Bloom's framework is *analysis.* Comprehension emphasizes the meaning and intent of the material. Application involves remembering while making the appropriate generalizations or principles about the material. Breaking down material into its constituent parts and detecting the relationships of the parts and the organization distinguishes analysis from application. Analysis also involves the techniques and devices used to convey the meaning and establish the conclusion of a communication. Educationally, analysis aids in full comprehension of the material. This learning level is comprised of three categories: elements, relationships, and organizational principles. Analysis of elements involves inferring unstated assumptions in a document and distinguishing facts from normative statements. After identifying the elements, the student must determine the relationships among the elements. The analysis level involves detection of parts of the thesis versus expansion on the thesis. At this level in accounting, students distinguish between expenses and liabilities (e.g., the difference between cash payments, deferrals, and accruals of expenses). The most complex component in this level is identification of organizational principles by analysis of a communicator's purpose, point of view, pattern of communication, or attitude, which may be relevant to evaluation, the sixth level in the process.

Bloom's Fifth Level: Synthesis

Analysis spills over into the fifth level in Bloom's hierarchy. *Synthesis* is the process of connecting elements to discern a pattern in the communication. This category in the cognitive domain affords creative behavior on the learner's part. This creativity is within some theoretical framework. Many educators mistake essay exams as providing experience in this synthesis stage; however, this is rarely the case. For instance, if the student merely generates a product that is substantially the same as the original material studied, despite some ordering and evaluation of the material, the task fails to require synthesis. An accounting application of this involves students solving problems that have formats similar to homework problems, perhaps with only number changes.

The products distinguish the different categories of synthesis. In the first, the product is a unique communication. Creative writing or presentation assignments are examples of unique communication tasks. Students could create a fairy tale or poem that is a metaphor for an accounting principle to accomplish this goal. In the second category of synthesis, the product is a proposed set of operations. An instructor could assign the development of an audit plan to examine students' ability to synthesize. The product of the third category is a set of abstract relations gleaned from an analysis of observed phenomena, initially not explicit, but requiring skills of deduction. To meet the goals of abstraction, students could complete a case in which a company known to have failed is analyzed for liquidity and solvency prior to the year of bankruptcy.

This important category relies on the students' personal participation and independence of thought and action, key to the concept of lifelong learning. As Bloom (1956, 167-68) stated:

> Especially important too are the tremendous motivational possibilities in synthesis activities. Such tasks can become highly absorbing, more so than the usual run of school assignments. They can offer rich personal satisfactions in creating something that is one's own. And they can challenge the student to do further work of a similar sort.

Students who can produce an audit plan, test theories, or formulate generalizations about a set of financial statements have satisfied the thinking skills necessary to synthesize.

Bloom's Sixth Level: Evaluation

Evaluation is the highest order in Bloom's taxonomy. It is defined as a learner's ability to use standards and criteria for making quantitative and qualitative judgments. Evaluation combines all of the other levels of Bloom's taxonomy. Although some value judgments may be involved, this stage is primarily cognitive as opposed to emotive. Evaluation is not always the last step in the thinking process; instead, it may precede new knowledge. Some judgments involve an evaluation of evidence such as logic, consistency, and internal criteria. Others are made by referencing external criteria, such as comparing several items based on standards within a field. For instance, in accounting, an investor can analyze two sets of financial statements (either of one firm across years or across firms in a single year) using a set of decision criteria constitutes evaluation. The set of criteria decision-makers use is not necessarily constant across decisions. Bloom recommends answering the following questions to determine the relevance of the criteria to the task: "Do the means employed represent a good solution to the problem posed by the ends desired? Are the means the most appropriate ones when the alternatives are considered? Do the means employed bring about ends other than those desired?" (1956, 191). The student also could compare theories, weigh values in alternate courses of action, and identify judgments when applying standards. An assignment in which the company has made several errors and has irregularities in its financial statements requires selection of an audit opinion. This exercise satisfies the requirements of evaluation: the student assesses materiality, determines conformity to GAAP, and employs professional judgment.

While synthesis and evaluation are at the highest levels in the taxonomy, the degree of difficulty does not necessarily increase as the student proceeds through the taxonomic levels. For instance, an instructor could ask a question about a simple income statement shown on a transparency. The activity requires the student to synthesize skills. Another question, which may seem trivial, ultimately is a more difficult question, despite its classification in Bloom's lowest level: on what date was APB Opinion #17 adopted?

A summary of Bloom's six levels of cognition appears in Exhibit 2. There are two categories of objectives in Bloom's cognitive domain, knowledge and intellectual skill. Each of the top five levels in the taxonomy requires higher-level thinking skills beyond rote-learning of facts.

TAXONOMY'S RELATIONSHIP TO CRITICAL THINKING

The intention of Bloom's taxonomy is to control instructor inputs, not student outputs. Accounting faculty can use techniques that incorporate Bloom's highest levels of cognitive development, helping students learn how to learn. Bloom's taxonomy provides a structured means to teach critical thinking. "The key to development of critical thinking skills is not what approach should be used but rather the degree to which a teaching approach is structured" (Riordan and St. Pierre 1992, 63).

The emphasis across disciplines needs to be on inquiry, abstract logical thinking, and problem solving skills. Since traditional college teaching practices have focused on transmitting knowledge rather than on developing thinking skills, more relevant teaching methods must be developed (Cunningham 1996, 50).

Thinking is defined as "an active, purposeful organized process that we use to make sense of the world" while critical thinking is defined as "making sense of our world by carefully examining the thinking process in order to clarify and improve our understanding" (Chaffee 1990, 1, 37). Thus, critical thinking is the process of evaluating a problem and synthesizing all information (Kimmel, 1995; Bierstaker, Bedard, and Biggs 1999). As Finocchiaro (1990, 465) said:

> I believe that it is probably true that all thinking which is reasoned *and* evaluative *and* self-reflective is critical thinking. Then insofar as reasoned, evaluative, and self-reflective are three senses of "critical," we may also say that critical thinking is, indeed, thinking which is critical.

Table 2.   Taxonomy of Educational Objectives in the Cognitive Domain

Knowledge	Intellectual Abilities & Skills
Knowledge-ability to recall information • terminology • specific facts • conventions • trends & sequences • classification • criteria • methodology • universals in a field • principles & generalizations • theories & structures	Comprehension-understanding • translation • interpretation • extrapolation
N/a	Application-use of abstractions in concrete situations
N/a	Analysis-ability to break down information into constituent parts • elements • relationships • organizational principles
N/a	Synthesis-put together parts to form a whole • produce unique • communication • produce plan of operations • derive set of abstract relations
N/a	Evaluation-judgments about value of material • internal evidence used • external criteria used

Accountants must be able to utilize critical thinking skills. Today's world requires the ability to synthesize new information continuously (NIE 1984, 43). Many accounting organizations, particularly educational committees, have espoused the position that learning should be active and continuous (i.e., lifelong) (e.g., AAA 1986; AECC 1990; Arthur Andersen et al. 1989). For example,

> Individuals seeking to be successful in the diverse world of public accounting must be able to use creative problem solving skills in a consultative process. They must be able to solve diverse and unstructured problems in unfamiliar settings. They must be able to comprehend an unfocused set of facts; identify and, if possible, anticipate problems; and find acceptable solutions (Arthur Andersen et al. 1989, 6).

In addition, accounting graduates should be able to communicate effectively in both written and spoken language. Instructors who use Bloom's taxonomy can achieve these educational objectives.

BLOOM'S SEEDS IN THE ACCOUNTING CLASSROOM

Too often students operate at the lowest levels of thinking in Bloom's taxonomy, *knowledge and comprehension* (which involves remembering material and grasping meaning). To maximize a course's benefits, the class discussions should be at the middle levels of thinking, *analysis and application* (using information and breaking it down into its parts). Exams should incorporate all levels in Bloom's taxonomy, and student papers should *synthesize* the information garnered from the readings, research, and classroom discussions, pushing learners into the highest level of thinking: *evaluation* (putting parts into the whole and judging the value of something for a specific purpose given definite criteria).

The purposes of writing in the accounting curriculum are many and varied; however, the common thread in all prior literature seems to be a shared goal of preparing students for success in the profession. The seven aims espoused in past research--learning to write, writing to learn, fostering critical thinking, sharpening organizational abilities, reflecting mastery of a body of information, synthesizing information, and analyzing the audience (Stocks, Stoddard, and Waters 1992, 196-197)—can be accomplished by using Bloom's paradigm. Exhibit 3 presents a selection of the many and varied types of writing utilized by accounting educators.

Some accounting literature suggests the use of writer-based prose (Wygal and Stout 1989; Hoff and Stout 1989; Cunningham 1991). Locke and Brazelton (1997) recommend attending to the process of developing the message rather than on the transmissions. Stocks, Stoddard, and Waters (1992) present a checklist of formats available to accounting faculty: major research papers, a problem written on a topic (and its accompanying solution), memos or letters addressed to a CFO, partner, manager, or controller, a one-paragraph summary of the major issues presented during the class discussion, a comprehensive question written on a point that is unclear to the student, a position paper, and articles written on current accounting developments for local or student newspapers. Corman (1986) advocates short format papers, such as letters, for ease of grading.

The process of sophisticated thinking generally requires writing because much of what is thought or said is lost--the mind can hold only so much at a given moment. Expansion of ideas requires repetitively returning to them, accomplished most easily through writing. To reach Bloom's higher cognitive levels, it is usually necessary to assign longer papers to allow development of ideas, although two pages may be adequate if the topic is sufficiently narrow.

An effective approach to incorporating Bloom's taxonomy into the accounting classroom is to treat each class period as a scholarly argument on an accounting topic. The class should begin with a question, the basis for most scholarly arguments. Students can discuss what they already know (the facts). Readings and homework assignments establish common knowledge. At this point, the instructor could clarify any misconceptions students hold. Then the class moves into the

Table 3.   Type of Writing Used in Accounting Literature

Accounting Authors	Type of Writing Advocated
Andrews and Pytlik (1983)	Letters to diverse audiences Evaluation of others' writing
Corman (1986)	Memos and letters
DeLespinasse (1985)	Memos and letters in response to client questions
Hirsch and Collins (1988)	Homework assignments, cases, and exam questions
Mohrweis (1991)	Memos to three different audiences
Stocks, Stoddard, and Waters (1992)	Major research papers, a problem, memos, in-class paragraphs, position paper, articles
Locke and Brazelton (1997)	Process of developing the message in which rewriting is emphasized
Scofield (1994)	Double-entry journals and contemporary readings
Scofield and Combes (1993)	Process-oriented assignments

uncertainties, discussing alternate views with supporting arguments. In many cases, the instructor can use examples for which there is no one right answer, but a plethora of alternatives to demonstrate the concept at each of Bloom's levels. Depending on the instructor's purpose, a common thread can hold a number of disjointed topics together during the class period. The decision the class ultimately makes is less important than the process because the effort ultimately translates into the ability to make a decision; but, "real life" decision-making requires choosing the *best* alternative rather than the only alternative.

Bloom's taxonomy, while explaining the levels of cognitive processing, formulates the theory of mastery learning. The instructor's role should be to help start the discussion through engaging readings and relevant examples of the subject matter. The instructor should lead the discussion in an open manner, without taking responsibility from the students. For example, the instructor should comment: "Can you say more about that?" "How do you react when *x* happens?" "Tell me more." The idea is to create an atmosphere where students feel able to expand their ideas and express thoughts even when they are not yet formulated. At that juncture, the instructor should respond with supportive statements such as: "I see;" "You're on a good track there;" "What an interesting concept."

If the purpose of a particular class session is to make a decision involving tax issues in a business environment, the instructor constructs facts that will fit with the topic of the "lecture." Then, the instructor begins a discourse covering the controlling statutes, regulations, and case law. Finally, with the assistance of class participants, the instructor extends the topical analysis by considering alternate positions also supported by existing law. All aspects of the discussion should support the idea that there are (1) a set of facts for a given case, (2) some rules that govern the decision-making process, (3) others' interpretations of the outcome in a given circumstance, (4) consideration of the variety of decisions that could be made, and (5) the decision with supporting documentation for that position. At any point, altering the facts to fit the topic that needs coverage is desirable, focusing on sprinkling it with many "what-ifs." This approach is even more effective with research assignments made outside of class or with homework problems because the student is able to delve into the topic at a depth that is difficult to achieve in the classroom.

Teaching students in steps helps them understand the cumulative nature of most learning. The ideal approach to teaching with Bloom's taxonomy is to lay the foundation with the first three levels (knowledge, comprehension, and application); then, spend most of the class time in the higher levels (analysis, synthesis, and evaluation). It is necessary to determine the level at which students are thinking and identify their weaknesses. An instructor can use pop-quizzes that provide questions at each of Bloom's levels or pose probing questions in the classroom to determine the students' levels in Bloom's taxonomy. A student may be good at higher levels and still be struggling at the lower levels. Another student might be frustrated at evaluation and synthesis levels if materials from the first three levels are missing. These principles apply to out-of-class assignments as well.

Instructors should make and grade assignments in steps since the progression through the levels assists with future assignments. As an illustration, students in a financial accounting course could select a Fortune 500 corporation from instructor-designated industries. In the initial assignment, in which the instructor grades only style and content, students could explore the company's beginnings at the knowledge and comprehension levels (Who started it? What kind of investment was required? What was the original mission? What was the company's line of business? Is it a different company now?). The second assignment of the semester would involve further research into the students' chosen businesses. Students would investigate current information on the companies, including accounting principles used, type of audit opinion received, media publications involving this company or industry, etc. The paper is graded on style, content, mechanics, and the student's ability to analyze company information (application and analysis levels). The final paper evaluates the company as a prospective investment, requiring financial statement analysis (synthesis and evaluation levels). Grouping selected companies by industry, with each student responsible for reporting to the industry groups the information about that particular company within the industry, also provides a team project experience. The team then synthesizes industry information and makes a presentation in which they draw conclusions regarding industry reporting standards, industry financial health using comparative ratios, and any other rational financial conclusions (synthesis and evaluation levels). This project

Table 4.   Applications of Bloom's Taxonomy to the Accounting Classroom

	Tax	Principles	Auditing
Knowledge	What is the economic definition of income? Define income under section 61. Define tax.	What is a debit? a credit? What is the fundamental accounting model? What is the definition of asset?	Define independent? What is the code of professional conduct? List the standards of field work.
Comprehension	What is the formula for computing tax liability? What are some sources of income? Explain to someone unfamiliar with an income tax system what it is.	What is an increase in an asset- a debit or a credit? Explain the fundamental accounting model in words. Give an example of an asset.	Provide an example of a generally accepted auditing standard. What is a confidence level?
Application	A single taxpaper's taxable income is $50,000, what is tax liability using the formula? Compute taxable income using standard deduction and a single exemption.	When would you debit an asset? If a buys an asset, what happens? Place the following accounts into their appropriate balance sheet category.	Which standard applies in situation x? Describe the responsibilities of an external auditor. What is the primary purpose of an operational audit?
Analysis	Distinguish between the economic definition of income and the definition of taxable income. Delineate exceptions to the broad definition of income. Identify sources of these exceptions to the §61 definition of income.	When would the payment of a liability not increase expenses? How does income end up in the company? Why can paying for something increase the company's expenses? Why doesn't a loan increase revenue?	Which audit opinion would you issue in siutation x? Does kiting exist in situation x? What are some alternatives to normal Accounts Receivable confirmation procedures?
Synthesis	Given the following 10 items of potential income, what is taxable income, providing reasons for inclusion or exclusion? How can this person save taxes?	Prepare financial statements. How do revenues and expenses end up on the balance sheet? Should the company invest in product X? Find dividends paid given the other components of retained earnings. Prepare a budget. Given company performance, propose a layoff of workers in a particular department.	How would you test to determine x? Plan the audit for the Accounts Payable for a software development corporation. Given the following errors/irregularities, is there a material irregularity as defined in Statements on Auditing Standards?

(continued)

Table 4   (Continued)

	Tax	Principles	Auditing
Evaluation	Explain why you think you chose the best approach for this taxpayer. Why might this not be the lowest tax? Go back to your decision with a critical eye and evaluate your synthesis are the parts legal, given the tenousness of this position, does this approach confirm to your personal professional value?	Given the following financial statements, compare them with the budget. Why is X the right answer? Why might it not be correct? Defend your answer using standards to make an argument. Based on the company's balance sheet, which company would you advice your client to invest in? Why?	Develop a risk evaluation model. Compute the estimated audit fee for company x. How effective is your strategy to assess internal control?

is an example of the stages of writing (and, ultimately, oral presentations) an instructor can use to achieve critical thinking. As Fulwiler (1987, 10) said:

> If we want writing (and thinking) skills to become useful, powerful tools among our students we must ask them to write (and think) in a context which demands some measure of *personal commitment*- which, in schools, is more likely to be in their major discipline than in specialized composition classes... Schools that offer most of their instruction through large classes, lectures, rote drills, and multiple-choice examinations obviously do little to nurture each student's individual voice. Schools that offer small classes, encourage student discussion, and assign frequent and serious compositions do nurture that voice.

The college educator's responsibility is to help students learn. Table 4 demonstrates the types of questions that an instructor could ask in an accounting classroom and represent each of the levels in Bloom's taxonomy. They achieve critical thinking goals in the classroom setting.

The goal of Bloom's taxonomy is to take students to the upper levels of thinking. Students usually are skilled at repeating facts but much less skilled at evaluating facts and intangible topics. Success in accounting careers requires Bloom's analysis, synthesis, and evaluation skills. Bloom (1956) provides a structure for teaching accounting students to be critical thinkers. His taxonomy also can teach creativity because the line between creativity and critical thinking is ill-defined and quite malleable. Construction of an argument overlaps the thought processes creative writers and critical thinkers use to make better critics after they write creatively. Assignments that require accounting students to turn an accounting situation or transaction into a fictitious scenario that is a metaphor for a "real-life" situation are effective tools for creative writing. This task requires that students understand the topic they are addressing and combines creativity and critical thinking skills. Students could write a fairy tale about an accountant (e.g., a day in the life of an accountant), an accounting transaction, a business situation, or a product (e.g., something as ridiculous as a pet rock) in an introductory accounting class. Asking them to present an accounting situation in a metaphorical way forces them to have a deeper understanding of the concept to which the metaphorical assignment pertains. The resulting discomfort many students experience in this task may stretch them to learn accounting at a higher level. The subsequent section illustrates methods available to accounting instructors designed to incorporate Bloom's highest levels in the classroom.

TWENTY WAYS TO ENCOURAGE CRITICAL THINKING IN THE ACCOUNTING CLASSROOM

Although not all of the 20 ideas suggested are in Bloom's sixth level, evaluation, all of the ideas are at the highest three levels. A summary of these 20 ideas for incorporating Bloom's taxonomy into the accounting classroom appears in Table 5.

Number 1: In-Class Free-Writes

Informal writing done during a class period can accomplish several goals. Students could write in the middle of a class period to refocus discussion or discern confusion on a particular issue. Periodically (e.g., every three or four weeks), an instructor could use the in-class free-write time (ranging from five to fifteen minutes) to determine the status of the class. Asking about lingering questions or inquiring about the most significant thing learned during the class period (or in the last three weeks) may uncover problem areas. Uncensored freewriting is another method that increases student thinking abilities. Some professors always ask a question (thereby providing feedback on specific issues and focus on a particular level in Bloom's taxonomy); however, the assignment just as easily could remain unfocused, simply asking the student to think aloud on paper (ask them to choose a topic: what was my biggest difficulty with the reading, homework, yesterday's test, etc.?). This informal writing always should be an ungraded assignment, except to consider it complete or incomplete.

More frequent writing in the style of the accounting profession also is useful. For instance, students could explain an accounting principle that was covered in class or that was part of their homework or reading assignment for the day. In a tax course, free-writes generating the hypothesized logic behind certain Internal Revenue Code provisions are effective. Evidence suggests that students who generate their own rationale for tax rules are better able to recall those rules (Schadewald and Limberg 1990; Hermanson 1997).

Table 5. Summary of 20 Ways to Encourage Critical
Thinking in the Accounting Classroom

1.	In-class free-writes
2.	Testing
3.	Explain errors on tests
4.	Writing test questions for class review
5.	Papers
6.	Rewriting
7.	Evaluating others' writing
8.	Journals
9.	Projects
10.	Class listserv
11.	Author an accounting text
12.	Class notes
13.	Editorials
14.	Divided-page journals
15.	Small groups
16.	Real-life cases
17.	Oral reports
18.	Class discussion
19.	Discussion preparation
20.	Get notes from a good student

Number 2: Testing

On tests it is imperative that professors include problems across the Bloom
levels so that students who are operating at the lower levels are able to complete some parts of the test. Although the goal is to get students to think at the
higher levels in Bloom's taxonomy, instructors need to offer students the
opportunity to work up through those levels. Some students are slower to reach
those levels than their classmates. Ideally, instructors should give students the
opportunity to demonstrate knowledge in Bloom's first two levels, representing roughly half of the exam. Then, instructors should ask fewer and fewer
questions at the highest levels in the taxonomy. This technique assures the
professor that students have achieved base-line knowledge and separates the
higher-order thinkers.

Instructors tend to ask easily-graded essay questions which are usually at
Bloom's lowest levels. It is crucial to students' cognitive development to include
problems that require synthesis and evaluation skills. Examples of simple essay
questions that require synthesis or evaluation include the social, political, or ethical contexts of particular pronouncements or positions taken in accounting scenarios. For instance, students could explain the political and ethical implications
of taxing certain income tax exclusions (e.g., social security or welfare payments).

Number 3: Explain Errors on Tests

Students are able to learn from their mistakes if their instructors require them to provide written explanations of errors on their tests, explaining where things went wrong and why. This process helps students find the missing link in their problem-solving process and identify their own weaknesses. The instructor models the importance of learning from mistakes by forcing students to recognize where their faulty logic occurs. This technique also places a priority on learning the material rather than on grades. In areas such as accounting, that initially develop in a spiral (starting at the bottom and picking up pieces at various levels), students must correct their errors before proceeding.

Number 4: Writing Test Questions for Class Review

Including a requirement to have students write their own test essay questions and word problems asks students to demonstrate an understanding of concepts in words and demands that students provide clear and specific instructions. If students share the questions or the instructor selects them, this exercise becomes a great test review that allows students to detect errors in their thinking and provides a break in the usual method of reviewing for exams.

Number 5: Papers

Term papers offer students an opportunity to think creatively and critically. Students could relate accounting to another discipline, consider the practical application (real-world) of some accounting principle, or devise a question and the ensuing scholarly argument (e.g., Why do accounting principles require ignoring human resources or the impact of inflation? Why not account for pensions as paid?). An effective assignment recommended by Wolcott and Lynch (1997), is "For whom do auditors work?"

When grading, the instructor should concentrate on the students' ability to analyze, synthesize and evaluate. If the instructor is a fanatic about the details (as accountant-types often are), then the instructor should correct mistakes only on the first page of the rough drafts, and students could correct similar mistakes in the balance of the paper in their final versions.

Number 6: Rewriting

Instructors should permit students to revise their papers because they learn the skills needed to make their writing better from feedback. If they have not communicated well, how does awarding them Ds or Fs on their papers serve them in the future? They are able to learn so much more by rewriting (if they choose) what they have been working on already. Most students do not refer back to a previous, unrelated paper to write subsequent papers; however, they must actually learn the skills needed for improvement in the process of rewriting.

Dossin (1997) uses the terms "finished writing" and "unfinished writing" to describe student papers. She claims that bad writing does not really exist; rather, some writing is simply unfinished, requiring additional rewrites. Learning is a process facilitated by the process of writing and rewriting. As Fulwiler (1987, 13) said:

> Every time you ask your students to revise a paper again before handing it in, you are asking them to *rethink* that paper. When you point out missing evidence, poor organization, or faulty logic, you ask writers to reconceive and rereason more carefully. When students are asked regularly to rethink their work, they will, in fact, learn to think more methodically, carefully, and convincingly, which is the intent of our instruction in the first place.

Number 7: Evaluating Others' Writing

Another effective technique in teaching students to be better writers is to provide samples of papers. At some point before the first formal writing assignment, the instructor should distribute two or three papers written during previous semesters (all on the same topic). The students should have a copy of the evaluation scale used for grading their papers and have the opportunity to discuss the standards during class. Students could grade several papers and bring them to class and discuss grading techniques. For example, the instructor could open the discussion with "Why is paper #1 a D while paper #2 is an A+?" "What worked for the writers?" or "Identify the thesis and determine whether the author successfully supported that thesis." This process of evaluation accomplishes the goal of higher-order thinking.

Number 8: Journals

Journals provide another effective means of forcing students to think critically, while ensuring class preparation. Student can use journals to summarize readings, class coverage, attitudinal perspectives, questions about readings or class material, or personal responses to any of the above. In a meta-analysis of approximately 500 studies, journal-writing provided superior learning to free-writing (Hillocks 1986). Periodic review of student journals also gives the instructor the opportunity to detect problem areas in student logic.

Number 9: Projects

Projects often require higher-order thinking; therefore, they make good assignments to accomplish the tasks in Bloom's taxonomy. Any idea that makes students think through an entire process is effective. For example, students could choose a particular accounting career. They could formulate a series of questions necessary to determine whether they would be interested in pursuing a career in that field of accounting. This assignment could be followed by role-playing inter-views with accountants to discern the faults in their logic, allowing classmates to demonstrate career knowledge and debunking stereotypes of certain accounting careers. Another stage in the project might include an actual interview of an accountant in that career.

Number 10: Class Listserv

Another idea for achieving higher-order cognitive skills utilizes a class listserv. The instructor could impose a requirement that each student posts relevant ques-tions or suggest potential responses at specific points during the semester. Writing the questions transfers students' confusion to "paper." In the process of articulat-ing a problem, the students have narrowed the area of uncertainty (perhaps making it less intimidating). At the same time, other students benefit from the confidence they gained by responding to classmates' questions.

Number 11: Author an Accounting Text

Montague (1973) incorporated an unusual approach in a math class, which is as effective in an accounting course. Some students had the opportunity to author a mathematics text. The first draft required peer review, followed by group discus-sion on the second draft. Simultaneously, other students acted as editors, which forced them to understand the material in depth. Students alternated writing particular sections of each chapter, as well as undertaking the roles of author and editor. By requiring research into mathematical theories and locating other texts on the topics, the task also fulfilled research requirements for the course.

Number 12: Class Notes

A variation on the previous approach is to have students take turns writing class notes formally and distributing them to their classmates. They ultimately create a lecture note collection rather than a book. This exercise achieves numerous bene-fits: it teaches responsibility, empowers students, requires synthesis of material, recognizes shared knowledge, and relies on peer cooperation and participation for information instead of solely on the instructor.

Number 13: Editorials

Students could write an editorial letter as an exercise. Although this is particu-larly good for ethics assignments, it as easily could apply to an intermediate accounting or theory class in which the students respond to an exposure draft of a proposed change in accounting standards.

As a variation of this approach, King (1982) used letter-writing as a means to describe to a friend how to solve a math problem or discuss feelings about or difficulties with certain concepts. This idea also could apply to accounting coursework.

Number 14: Divided-page Journal

An excellent idea for instructors who assign homework and periodically collect it is the divided-page homework journal. On one-half of the page, problems, solutions, and calculations appear. On the other half, students give written explanations of their work, including any problems encountered (Why is the problem difficult? Where did I get stuck? How could I make this problem easier?). This assignment fosters a sense of ownership of the students' own learning.

Number 15: Small Groups

One approach designed to encourage students to operate at the higher levels in the taxonomy is the use of cooperative learning. Students work in small groups with two to five classmates to solve a problem that requires higher-level thinking. The entire group is responsible for making certain that each member of the group understands and can present the information to the instructor. This understanding is accomplished by having the group members take turns asking each other questions that the instructor might pose during the subsequent whole-class discussion of the problem. Therefore, the task of building on Bloom's six levels falls on the group members (Cottell and Millis 1993; Clarke 1989).

Number 16: Real-life Cases

Scientists have utilized a technique that could be equally effective in accounting classes. Instructors assigned real-life, open-ended, mini cases (Smith et. al. 1985). For example, participants received the task of creating a small expert system that would choose the best rock-blasting technique for mineral mining. The result was that students learned the topic much better than from the standard lecture approach. Instructors should require these learners to have a complete understanding of the theories and facts relevant to the project. Because of the "if-then" approach utilized, the project demanded synthesis and evaluation of all relevant information. In accounting, students could create an expert system that selects an acceptable level of material error, of audit risk, or uncertainty in tax cases.

Number 17: Oral Reports

Yet another idea on this same theme is one that is familiar to instructors. Until someone teaches a concept to another, there is uncertainty about the level of understanding. All of one's errors in assumptions and logic become apparent in a teaching environment. The accounting student could prepare an oral report on the homework, on the reading, or on a current event as it relates to the class topic of that particular period. This exercise serves to reinforce understanding of the class material.

Number 18: Class Discussion

Barnes (1983) reported that only 3.65% of class time is devoted to faculty questioning students. The study found no statistically significant differences across fields. One unexpected result was that instructors spent even less time questioning in upper-level courses where the students have all of the fundamentals from the lower levels of thinking and should be learning to synthesize and evaluate. Of those questions asked in the classroom, only 5% fell into Barnes' two categories most closely related to Bloom's top levels: *divergent* (questions that ask for responses that elaborate, implicate, and synthesize) and *evaluative* (questions that require students to rate, judge, and qualify).

> Our sense of mythology suggests that in college one would expect to find inquiring young minds being challenged by the intellectual and perceptive questions of learned professors... (However), many of the classes are void of intellectual interchange between professors and students. Making instructors aware of how they contribute to the problem is a large part of the solution. Few in our ranks would argue that on most days, more than 4% of the class time could be devoted to questioning (Weimer, 1996, 70).

Discussion-oriented classes are more effective than lecture methods because they force students to operate at the higher levels in Blooms' taxonomy (McKeachie 1978). Further, McKeachie asserts that material learned in discussion results in retention superior to that achieved in lecture format and that students also prefer discussion courses to lecture courses. Although it is most effective to pose broad open-ended questions with no "right" answer, the topic must be well-defined. Beginning the class discussion with a common experience (generally from the readings) accomplishes a discussion-oriented class followed by an analysis of a specific problem (with the instructor playing the role of benign disrupter). Prior to class, the instructor should prepare a questioning strategy (i.e., plan a sequence of questions designed to achieve a teaching goal).

Teaching methods in the accounting classroom that continually incorporate discussion demonstrate effective techniques for the learners to employ in higher-order thinking processes. During the first class period of a course, an instructor might consider talking about the stereotype of an accountant or a business executive. Posing this type of broad, open-ended question early in a course stimulates thought and sets the mood for future discussions. Other class periods should begin with a concrete example, experience, or news article designed to generate discussion.

Number 19: Discussion Preparation

Students must prepare and send electronic responses to the instructor's five questions relating to the next period's assignment (via e-mail, a class listserv, or a web page). These communiques allow the instructor to anticipate problem areas while allowing students to be fully prepared to participate in a detailed discussion on a topic. Alternately, if an electronic medium is unavailable, students could pose questions about the textual reading or homework problems, achieving similar class preparation goals.

Number 20: Get Notes from a Good Student

Every week or two, the instructor should collect notes from a conscientious student. This allows the instructor to know what is really getting across, particularly the level in Bloom's taxonomy at which the classroom discussion is occurring. The student would have to agree to this at the beginning of the semester; likewise, the instructor could alternate the student participant.

CONCLUSION

Accounting students should become critical thinkers, able to synthesize a wealth of information. The ability to think critically and creatively is of tantamount importance to success in the accounting profession (AECC 1990). Using a variety of pedagogues, accounting instructors ought to lead their students from Bloom's lowest level of learning (fact assimilation) to the highest levels of learning (synthesis and evaluation). The classroom should act as an intellectual community that promotes development of cognitive abilities. Practitioners constantly are synthesizing information, from analyzing internal controls and measuring materiality of a financial statement error during an audit, to selecting the "right" solution to a client's tax issue.

Discussion pushes students beyond their prior knowledge into new knowledge (from their perspective) that they can understand. During discussions, students produce information, not only evaluate and synthesize material. We must move our accounting majors from the role of *consumers* of information into *producers* of information. Instructors must change from imparters of information into facilitators of discussion.

We should foster learning by giving students more responsibility for the learning that occurs in the classroom. We accomplish student ownership of accounting material in an interactive learning environment. Talking or writing about material leads to better learning than the traditional lecture format. Also, making the students responsible for probing issues and debating with each other during class yields more critical thinking during discussions, removing some of the pressure to make more written assignments that achieve higher-order cognition. While the class discussion may be geared to the middle-range student, writing assignments can force even the brightest students to expand their intellectual development. Assignments that encourage creative, critical, and original thinking should be our goal.

The entire accounting faculty must accept this challenge--the challenge of producing graduates who are competent critical thinkers and writers. If students taking sequentially-ordered courses are not exposed to communication at higher cognitive levels throughout the curriculum, those skills learned in the early accounting courses never are developed fully. Likewise, if not all students taking a particular course (i.e., not all instructors teaching different sections of the same course) are exposed to the same demands, it is difficult to build on the skills learned in those earlier courses. Programmatic commitment and priority offer more benefits than single-course use; but, any exposure to critical thinking is positive for students.

> While significant progress is being made in articulating and promoting this overall theoretical approach, there is currently no consensus about the efficacy of specific classroom strategies, due to at least three reasons. First, many of the commonly recognized models are inadequate for use in critical thinking development. Second, there are few techniques that allow professors to assess their students' critical thinking efficiently. Third, few researchers have examined systematically the impact of specific classroom techniques on students' critical thinking (Wolcott and Lynch 1997, 61).

I am optimistic about the likelihood of success of this philosophy since Bloom's taxonomy has proved effective in other disciplines (Dossin 1997; Montague 1973; King 1982; Smith et al. 1985). There is reason to expect that the approach would flourish in accounting classrooms in which educational goals match those espoused by the accounting profession because critical thinking and communication skills are important. Widespread application of this methodology in accounting classrooms could benefit students, faculty, and the profession. Students may gain skills needed for success in the profession while enjoying a more active role in the learning process. A process that makes the classroom time more interesting may intellectually stimulate instructors. This approach eases the instructor's burden to impart knowledge, instead placing responsibility for learning on the students. The profession, in turn, gets workers who are well-prepared to meet the challenges of their chosen career.

Student-centered classrooms that utilize cooperative learning techniques are more effective than the traditional lecture format (Cottell & Millis, 1993). Traditional methods of evaluating accounting students still dominate our accounting programs. There has been increased integration of less traditional assignments such as cases and writing assignments; however, most faculty remain on the traditional paths (Gabriel and Hirsch 1992; Dow and Feldman 1997).

Critical thinking is vital to the future success of our students. Bloom's taxonomy provides a convenient framework with which to structure our curricula;

however, the particular model of cognitive development we use to guide our process is less important than accomplishing our educational goals. The sequencing of discussion and assignments directly affects students' intellectual development. Bloom's taxonomy allows us to categorize what our students have mastered. Spear (1983, 51) addresses the taxonomy:

> In this respect it provides a way to understand not what people know, but how they know it, a sequential epistemology in which mastery of the individual stages is equally as important as attaining the final stage.

This paper increases the canon of accounting education literature by providing a framework in which to structure teaching to focus on critical thinking skills and intellectual discourse. We can teach our students to write, use writing to learn, foster creativity, and emphasize scholarly debate with this framework. Our development as teachers is dependent on our ability to achieve educational goals that reflect the demands of a dynamic accounting profession.

REFERENCES

Accounting Education Change Commission (AECC). 1990. Position Statement No. 1: Objectives of education for accountants. *Issues in Accounting Education* (Fall): 307-312.

American Accounting Association Committee on the Future Structure, Content, and Scope of Accounting Education. 1986. Future accounting education: Preparing for the expanding profession. *Issues in Accounting Education* 1, 160-195.

American Accounting Association. 1968. *Guide to Accounting Instruction: Concepts and Practices.* Cincinnati, OH: Southwestern Publishing.

American Association for the Advancement of the Humanities. 1982. Anaysis: Schools. *Humanities Report* (February).

American Institute of Certified Public Accountants. 1988. *Education Requirements for Entry into the Accounting Profession.* New York: American Institute of Certified Public Accountants.

Anderson, L.W., and L.A. Sosniak, eds. 1994. *Bloom's Taxonomy: a Forty-year Retrospective.* Chicago: University of Chicago Press.

Andrews, J., and R. Pytilk. 1983. Revision techniques for accountants means more and efficient written communication. *Issues in Accounting Education*, 152-163.

Arthur Andersen & Co. et al. 1989. *Perspectives on Education: Capabilities for Success in the Accounting Profession.* New York: Arthur Andersen & Co., Arthur Young, Coopers & Lybrand, Deloitte Haskings & Sells, Ernst & Whinney, Peat Marwick Main & Co., Price Waterhouse, and Touche Ross.

Barnes, C. 1983. Questioning in the college classrooms. Pp. 61-81 in *Studies of College Teaching: Experimental Results, Theoretical Interpretations, and New Perspectives*, edited by C. Ellner and C. Barns. Lexington, MA: Lexington Books.

Bierstaker, J., J. Bedard, and S. Biggs. In Press. Fostering clinical thinking in accounting education: Implications of analytical procedures research. In *Advances in Accounting Education*, Vol. 2, edited by B.N. Schwartz and J.E. Ketz. Stamford, CT: JAI Press.

Bloom, B., ed. 1956. *Taxonomy of Education Objectives.* New York: David McKay.

Chaffee, J. 1990. *Thinking Critically.* Boston: Houghton Mifflin.

Civikly, J. 1986. Instructor communication habits: Confrontation and challenge. Pp. 5-9 in *Communicating in College Classrooms*. New Directions for Teaching and Learning, 26. San Francisco: Jossey-Bass.

Clarke, J. 1989. Designing discussions as group inquiry. *Classroom Communication*, 73-87.

Connolly, P., and T. Vilardi, eds. 1989. *Writing to Learn Mathematics and Science*. New York: Teachers College Press.

Corman, E. 1986. A writing program for accounting courses. *Journal of Accounting Education* 4(2): 85-95.

Cottell, P., and B. Millis. 1993. Cooperative learning structures in the instruction of accounting. *Issues in Accounting Education* 8(1): 40-59.

Cunningham, B.M. 1991. Classroom research and experiential learning: Three successful experiences—The impact of student writing in learning accounting. *Community/Junior College Quarterly of Research and Practice* (July-September): 317-325.

Cunningham, B. 1996. How to restructure an accounting course to enhance creative and critical thinking. *Accounting Education: A Journal of Theory, Practice and Research* 1(1): 49-66.

DeLespinasse, D. 1985. Writing letters to clients: Connecting textbook problems and the real world. *Journal of Accounting Education* 3(1): 197-200.

Dossin, M. 1997. Writing across the curriculum: Lessons from a writing teacher. *College Teaching* 45(1): 14-15.

Dow, K., and D. Feldman. 1997. Current approaches to teaching intermediate accounting. *Issues in Accounting Education* 12(1): 61-75.

Estes, R. 1979. The profession's changing horizons: A survey of practitioners on the present and future importance of selected knowledge and skills. *The Internal Journal of Accounting Education and Research* (Spring): 47-70.

Finocchiaro, M. 1990. Critical thinking and thinking critically: Responses to Siegel. *Philosophy of the Social Sciences* (December): 462-265.

Francis, M., T. Muldur, and J. Stark. 1995. *International Learning: A Process for Learning to Learn in the Accounting Curriculum*. Sarasota, FL: Accounting Education Change Commission and American Accounting Association Educational Research Committee.

Freire, P. 1970. *Pedagogy of the Oppressed*. New York: Herder and Herder.

Fulwiler, T. 1987. *Teaching with Writing*. Portsmouth, NH: Boyton/Cook Publishers.

Gabriel, S., and M. Hirsch. 1992. Critical thinking and communication skill: Integration and implementation issues. *Journal of Accounting Education* 10(2): 243-270.

Gebe, B. 1990. The grandaddy: Bloom's taxonomy. *Training: The Magazine of Human Resources Development* 27(3): 107.

Henry, L.G., and N.Y. Razzouck. 1988. The CPA's perception of accounting education: Implications for curriculum development. *The Accounting Educators' Journal* (Spring): 105-117.

Hermanson, D. 1997. The effect of self-generated elaboration on students' recall of tax and accounting material: Further evidence. *Issues in Accounting Education* 9(2): 301-318.

Hillocks, G., Jr. 1986. *Research on Written Composition: New Directions for Teaching*. Urbana, IL: ERIC Clearinghouse on Reading and Communication Skills and the National Conference on Research in English.

Hirsch, L., and J. Collins. 1988. An integrated approach to communication skills in an accounting curriculum. *Journal of Accounting Education* 6: 15-31.

Hoff, K.T., and D.E. Stout. 1989/90. Practical accounting/English collaboration to improve student writing skills: The use of informal journals and the diagnostic reading technique. *The Accounting Educators' Journal* (Winter): 83-96.

Kimmel, P. 1995. A framework for incorporating critical thinking into accounting education. *Journal of Accounting Education* 13(3): 229-318.

King, B. 1982. Using writing in the mathematics class: Theory and practice. In *New Directions for Teaching and Learning: Teaching Writing in all Disciplines*, edited by C.W. Griffin. San Francisco: Jossey-Bass.

Kohlberg, L. 1970. Education for justice: A modern statement of the Platonic view. In *Moral Education: Five Lectures*, edited by J.M. Gustafson et al. Cambridge, MA: Harvard University Press.

Kurfiss, J. 1989. *Critical Thinking: Theory, Research, Practice, and Possibilities.* Washington, DC: Association for the Study of Higher Education.

Laufer, D., and R. Crosser. 1990. The "writing-across-the-curriculum" concept in accounting and tax courses. *Journal of Education for Business* (November/December): 83-87.

Locke, K., and J. Brazelton. 1997. What doe we ask them to write, or whose writing is it, anyway? *Journal of Management Education* 21(1): 44-57.

May, G.S., and C. Arevalo. 1983. Integrating effective writing skills in the accounting curriculum. *Journal of Accounting Education* 1(1): 119-126.

McKeachie, W. 1978. *Organizing Effective Discussion. Teaching Tips: A Guidebook for the Beginning College Teacher.* Lexington, MA: D.C. Heath.

Mohrweis, L. 1991. The impact of writing assignments on accounting students writing skills. *Journal of Accounting Education* 9: 309-325.

Montague, H. 1973. Let your students write a book. *Mathematics Teacher* 79: 461-465.

Myers, I., and K. Briggs. 1976. *The Myers-Briggs Type Indicator.* Palo Alto, CA: Consulting Psychologists Press.

National Institute to Education (NIE) Study Group on the Conditions of Excellence in American Higher Education. 1984. *Involvement in Learning: Realizing the Potential of American Higher Education.* Washington, DC: NIE.

Nelson, I. 1995. What's new about accounting education change? A historical perspective on the change movement. *Accounting Horizons* (December): 62-75.

Novin, A.M., and M.A. Pearson. 1989. Non-accounting-knowledge qualifications for entry-level public accountants. *The Ohio CPA Journal* (Winter): 12-17.

Perry, W.G. 1970a. *Forms of Intellectual and Ethical Development in the College Years: A Scheme.* New York: Holt, Rinehart, and Winston.

Perry W.G. 1970b. *Forms of Intellectual and Ethical Behavior in the College Years.* New York: Holt, Rinehart, and Winston.

Phillips, M.J., and C.H. Davis. 1991. Writing requirements in financial accounting courses. *Journal of Education for Business* (January/February): 144-146.

Piaget, J. 1972. *The Principles of Genetic Epistemology.* New York: Basic Books.

Riodan, M., and E. St. Pierre. 1992. The development of critical thinking. *Management Accounting* 73(8): 63.

Schadewald, M., and S. Limberg. 1990. Instructor-provided versus student-generated explanations of tax rules: Effect on recall. *Issues in Accounting Education* 5(1): 30-40.

Scofield, B. 1994. Double entry journals: Writer-based prose in the intermediate accounting curriculum. *Issues in Accounting Education* 9: 330-352.

Scofield, B., and I. Combes. 1993. Designing and managing meaningful writing assignments. *Issues in Accounting Education* 8: 71-85.

Smith, K.A., A.M. Starfield, and R. Macneal. 1985. Constructing knowledge bases: A methodology for learning how to synthesize. *Frontiers in Education Conference Proceedings.*

Stocks, K.D., T.D. Stoddard, and M.L. Waters. 1992. Writing in the accounting curriculum: Guidelines for professors. *Issues in Accounting Education* 7(2): 193-204.

Weimer, M. 1996. Research summary: Professors part of the problem? *Classroom Communication,* 67-71.

Wolcott, S., and C. Lynch. 1997. Critical thinking in the accounting classroom: A reflective judgment developmental process perspective. *Accounting Education: A Journal of Theory, Practice and Research* 2: 1, 59-78.

Wygal, D.E., and D.E. Stout. 1989. Incorporating writing techniques in the accounting classroom: Experience in financial, managerial and cost courses. *Journal of Accounting Educaton* (Fall): 245-252.

AN EXPLORATORY EXAMINATION OF THE STUDY TIME GAP:
STUDENTS' AND INSTRUCTORS' ESTIMATION OF REQUIRED STUDY TIME

Robert H. Sanborn, Bill N. Schwartz, and
W. Darrell Walden

ABSTRACT

Time-on-task, or the amount of time students study, is a very important factor in the learning process. Most instructors are aware of the importance of time-on-task and attempt to communicate the importance of study effort to their students. However, anecdotal evidence suggests that most instructors believe students do not study as much as needed to master a subject, so the amount of study time becomes a point of contention. Instructors expect and tell students to study more but often students fail to live up to these expectations. Or do they?

We asked junior and senior accounting majors at five southeastern universities how many hours they studied for a specific course during the previous week and how many hours they believed their instructor expected them to study. We also asked

Advances in Accounting Education, Volume 2, pages 87-111.
Copyright © 2000 by JAI Press Inc.
ISBN: 0-7623-0515-0

their instructors how many hours they believed their students studied and how many hours they expected their students to study to master the course work.

We found that a gap exists between the amount of time students reported they study and the amount of time their instructors expected them to study. We also found that students study more than their instructors believe they study. Students believe that their instructors expect them to study more than their instructors expected them to study. Further investigation of these gaps between students and their instructors revealed that students who had instructors who expected their students to study more actually did report studying more. Students who had instructors, who expected them to study less, studied less. The following paper presents the results of our research and provides suggestions that may help decrease the study time gap that appears to exist (in student behavior).

INTRODUCTION

Learning is a complex and multi-faceted process. Students can learn from their activities in class (e.g., listening to lectures, participating in class discussions, working on cooperative learning exercises). They also can learn a great deal studying outside of class (e.g., reading the textbook or other course related materials, doing individual homework assignments such as textbook problems, writing research papers, participating in study groups and working on group projects). However, anecdotal evidence indicates that most accounting instructors believe students do not study enough outside of class. Conversations at educational conferences and in faculty lounges often will find some faculty members saying their students are not motivated and that they need to study more outside of class to master the course materials. To date little published empirical work in the accounting education literature supports this often-voiced concern.

Ultimately, instructors should want their students to perform to the best of their abilities. A gap between students' and instructors' estimations of required study time outside of class is dysfunctional toward that end. The current research study takes a first step by trying to document whether a gap exists. It is important that instructors know if a gap exists so they can find more effective means of communicating their thoughts about required study time.

We examine how many hours per week outside of class students report that they study the class materials of a specific accounting course and how many hours they believe their instructors expect them to study for the course. We also asked the students' instructors how much they thought their students studied and what they thought the amount of time per week their students needed to spend studying.

In the next section we discuss prior research about student study time and then present a *study time gap model* in the third section. The fourth section contains our hypotheses and in the fifth section we describe the methods we used to collect survey data. We discuss the results of our tests of the hypotheses in the sixth section. The paper ends with a discussion of the implications of our findings.

BACKGROUND

A few time-on-task studies exist in the education literature. In a study of 222 Cornell University students, Frisbee (1984, 345) found "that the time students allocate to a course does have a positive effect on course grade." He found that an additional 10 hours of study effort per semester had approximately the same impact on course grade as an additional 100 points on either the SAT- Math or SAT- Verbal scores. His findings strongly support the value of time-on-task.

Study habits and time-on-task have been paired in other research studies. Britton and Tesser (1991, 405) found that time management practices were better GPA predictors than SAT scores. Thombs (1995, 280) surveyed 576 freshman and found study habits and time-management skills to be accurate variables in discriminating between end-of-semester grade point averages. These findings showed that increased study time led directly to increased grades.

In the accounting area two studies found connections between student reported effort and success.[1] Ibrahim (1989, 57) examined the relationship between effort-expectation and students' academic performance in managerial cost accounting. His results indicate that students' effort levels have a significant relationship with actual performance. Parry (1990, 222) found that study group assignment may be detrimental to the student's effort if group study is viewed as a substitute for standard studying practices.

The question remains; why do students not recognize that increased study time should improve their performance? Other factors must have an effect on this behavior. Gleason and Walstad (1988, 315) failed to find that college students identified and used study time to maximize achievement. Their findings indicate that students were aware of the need to study more but failed to act on that knowledge in any significant way. Possibly faculty lounge cynics are right, and students are lazy. However, there may be other explanations: e.g. students may not know how much to study.

Kelly (1992, 3) found that beginning students had a high level of optimism about their classes but an unrealistic idea of the amount of study time required. Her later longitudinal research (1992, 18) showed continued student interest in the classes but increased problems with study time requirements. Students enjoyed their classes but never learned how much time to spend studying for them. Kember et al. (1996, 347) found a relationship between students' perceptions of workload and motivation. He determined that motivation improved when the workloads instructors required of students were realistic. Thus students might spend more time studying if they know that the study time goals are obtainable.

In summary, prior research supports the importance of time-on-task. The research shows a strong relationship between grades and the time spent studying. Students' study patterns and their motivation to study also impact their grades and time-on-task. However, no published research has investigated how much time students study outside of class for a specific course or have correlated that time with their instructors' perceptions. Our study examines whether there are gaps between student and instructor expectations concerning study time.

THEORETICAL FRAMEWORK

Lawler's (1967, 122) model of performance has been adapted for educational research by Ott et al. (1988, 131), Ott (1988, 378) and Parry (1990). The model presents performance as a function of an individual's effort, ability and role perception. It postulates that the effect and probability of rewards on performance, reward in this case being class grade, increase effort.

This research is an investigation of the "effort" component of the Lawler and Porter model. The probability that "reward depends upon effort" is the factor that underlies the accounting instructor complaint that students do not study enough. The model presumes that if students study more outside of class they will learn more. Therefore, we substitute "study hours" and "time spent studying" for "effort."

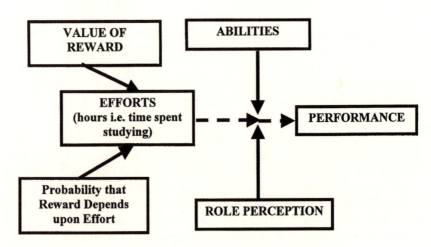

Source: Adapted by Parry (1990, 224) from Lawler and Porter (1967, 125).

Figure 1. Lawler and Porter Theoretical Model and Operationalization of Variable

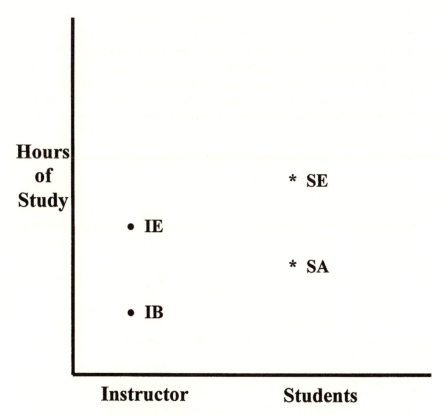

Hours of Study

* SE

• IE

* SA

• IB

Instructor   **Students**

Figure 2.

We propose a study time gap model component which depicts the expected relationships between instructors' perceptions of their students' study habits; instructors' estimates of optimal study behavior; students' actual study time; and amount of time students believe they should study. The *study time gap model* denotes IB as the number of hours per week that instructors believe their students actually study for the course. SA is the number of hours per week that students actually study for the course. IE designates the number of hours per week instructors expect their students to study for the course. SB is the number of hours per week that students believe they should study for the course.

We constructed the model under the assumption that instructors believe (1) their students study less than students actually study (IB < SA), and (2) that their students need to study more (IE > IB). These assumptions are consistent with the anecdotal evidence that many instructors are cynical about their students' study habits or that they underestimate the time it takes their students to master their lessons.

The model also assumes that students overestimate the time their instructors believe they should study (SB > IE). We believe that students overestimate the time they must spend to complete projects (Kelly 1992, 5). The model also indicates the belief that students spend less time studying than they believe their instructors expect them to study (SA < IE).

Finally, the model proposes that students believe they need to spend more time studying (SB > SA). The amount of time difference is important in this gap. If students overestimate the time needed to study for a course, they may be building motivational blocks into their perceptions (i.e., they may decide that they cannot possibly study as much as they must to succeed so they decide to stop studying at all or study less than they would with positive motivation). Significant time differentials may cause students to abandon their efforts instead of striving for an achievable goal.

HYPOTHESES

The model proposes that there are differences between what amount of time instructors think their students should study and reported student study hours, with instructor expectations being greater. The *study time gap model* (see Figure 2) suggests five specific study time gaps, which lead to the development of five hypotheses. These hypotheses, in alternative form, appear below.

First, the model proposes that instructors believe they can estimate the time that students study for their accounting courses, but accounting students study more than their instructors realize. Differences exist, which result in a *study gap*. We formulate the following hypothesis:

H1:   Accounting students report (SR) they spend more time studying than accounting instructors believe (IB) their students study (i.e., SR > IB).

Second, differences often exist between student study expectations and instructor study expectations. We believe that student expectations exceed what instructors expect. These differences result in a *study expectation gap*. We formulate the following hypothesis:

H2:   Accounting students believe (SB) they should study more than accounting instructors expect (IE) them to study (i.e., SB > IE).

Third, we believe instructors can motivate students. Unfortunately, too often these attempts to motivate students backfire and students actually become discouraged if they believe their instructors expect too much of them. As such, we suggest that students study less than they believe they should, which results in a *motivation gap*. We formulate the following hypothesis:

H3: Accounting students believe (SB) they should study more than they report (SR) they study (i.e., SB > SR).

Fourth, we suggest that accounting instructors expect their students to study more than their students report they study. Accounting instructors usually expect their students to study more than they do. This difference results in a *performance gap* between the instructor's study expectations and their students' reported study time. We formulate the following hypothesis:

H4: Accounting instructors expect (IE) students to study more than students report (SR) they study (i.e., IE > SR).

Finally, instructors tend to be cynical about their students' study habits. As such, we suggest that accounting instructors expect their students to study more than they believe their students study. The result is a *cynicism gap* between the instructor's expectation and the instructor's belief of how long their students study. We formulate the following hypothesis:

H5: Accounting instructors expect (IE) students to study more than accounting instructors believe (IB) their students study (i.e., IE > IB).

METHODOLOGY

We surveyed students to determine their study habits. Reed et al. (1984, 1031) conducted studies confirming the use of "Student Study Time Surveys," so we believe the self-reported survey technique would provide usable data. Independent volunteers administered the survey during class time for students (see Appendix A) and the companion survey for instructors (see Appendix B) in junior and senior level accounting courses at five southeastern universities. This use of an independent administrator assured students that their instructor would not see their responses, and we therefore believe their responses would fairly reflect their effort. The instructors completed their surveys out of the class room at the same time the students were completing their surveys. We believe this procedure insured that both students and their instructors were assessing the prior week's study time and that both parties were aware of the confidentiality of their responses. These universities differ in several respects (1) private and public, (2) rural and urban, and (3) entrance requirements, so we tested for differences in responses among schools. We found no significant differences.

We conducted the research during a week that neither was immediately before nor after an exam or major project was due. In short, we tried to choose a "normal" week. We conducted the survey for both students and faculty on either Monday or Tuesday of the week following the week upon which we asked students to report their study time. We believe this procedure captured the most accurate self-reported data possible.

We asked students to tell us the number of hours they studied for the course during the past week, hours they believed their instructor expected them to study, and hours they spent studying for all courses in the past week. We also gathered demographic data as to gender, age, marital status, year in school, overall GPA, number of hours gainfully employed, credit hours and number of courses currently being pursued, and ethnic background. In addition, we asked students about any unusual circumstances that might have affected their study time in this course during the current week. Only a few students reported anything unusual, and we did not consider these reasons extraordinary.

We asked the instructors for their estimates of how many hours their students studied for their course in the past week, and how many hours they believed their students should have studied. Instructors know that students come into a class with many demands on their time (e.g., family, friends, work responsibilities, other courses). However, we have noted that virtually every instructor we know gives their students some general guidance about how much time students will need to study for the class. Some students may study less or may study more depending a host of variables such as their performance goals, time priorities, how efficient they are with their study time, the environment in which they study, etc. Students decide how much time they devote to a class. The instructor's belief about how much time a student should devote to class is not based upon these individual factors, but rather how much time is needed to master the material. Our survey attempted to determine both the students' and instructors' beliefs and the students' actions. The instructors also indicated that there were no unusual events in their classes during the week in question that would have effected the amount of time the students studied for the course.

The survey asked students to report the time they studied for the course. The data collected from this question is not the same as the actual study time students spent (collecting of such detailed information would require a controlled lab experiment or other elaborate measures), but this self-reported information does indicate the amount of time students believe they study. We feel the self-reported time is a good surrogate for "actual study time" particularly since we surveyed the students so soon after the time period in question, and because the use of self-reported data is consistent with other studies cited above (i.e., students' reported study time [SR] is approximately equal to students' actual study time [SA]). "Study" was the variable we chose because we did not wish to differentiate between all possible study methods. In this initial research we just wanted to determine the existence of study gaps.

We collected survey data for one "normal" week rather than ask students to maintain a log for several weeks. We made this choice because we were concerned that a log would sensitize the students to the experiment so they would not act in a typical fashion. Adair et al. (1987, 5) provides procedures to control for the Hawthorne effect but these procedures would have limited the test population size because of the paired survey techniques required. We decided surveys were the best data collection instuments for this exploratory research. We also were concerned with surveying the students more than once in a semester. We felt it would be difficult to obtain data from two similar "normal" weeks and were uncertain if the students would change their behavior because of the first survey.

Table 1. Faculty Demographic Data

Demographic	Instructor Frequency	Percent	Section Frequency	Percent
Faculty Sex				
Female	4	33%	6	33%
Male	8	67	12	67
Total	12	100%	18	100%
Faculty Rank				
Assistant Professor	5	42%	7	39%
Associate Professor	5	42	8	44
Full Professor	2	16	3	17
Total	12	100%	18	100%
Years Teaching Course				
6 years or less	6	50%	8	44%
More than 6 years	6	50	10	56
Total	12	100%	18	100%
Years at University				
9 years or less	6	50%	8	44%
More than 9 years	6	50	10	56
Total	12	100%	18	100%
Course Level				
Junior Level	6	50%	12	67%
Senior Level	6	50	6	33
Total	12	100%	18	100%
Entrance Requirements				
Very Difficult	5	42%	9	50%
Moderately or Minimally	7	58	9	50
Total	12	100%	18	100%
Ethnicity				
White	11	92	17	94%
Minority	1	8	1	6
Total	12	100	18	100%

RESULTS

The results of our survey included responses from 363 students in junior and senior level accounting courses at five southeastern universities. Two universities were classified as having "very difficult" entrance requirements, two were classified "moderately difficult" and one "minimally difficult" by *Peterson's Guide to Four Year Colleges* (1999, 59-63). A total of 12 instructors who taught 18 class sections participated in the study. Twelve class sections were junior level accounting courses, while the remaining six were senior level accounting courses. After careful review of the 363 student surveys, we eliminated 22 student surveys due to student reported unusual circumstances occurring during the reported study week.[2]

Table 1 presents the faculty demographic data. There were four female and eight male instructors surveyed across 18 class sections. Five were at the faculty rank of assistant professor and seven were at the rank of associate or full professor. Half of the instructors had six years or less experience teaching the course. Fifty percent of the instructors had nine years or less experience at their present university. Six instructors taught junior level courses. Only one of the instructors surveyed was a minority.

In Table 2, demographic data of the students surveyed show that 53% were female, and 47% were male. Most students, 74%, were 21 years or younger, and 89% were not married. Minorities represented 20% of students. Forty-nine percent of the students were gainfully employed, 28% worked from one to 19 hours per week while 23% reported working 20 hours or more per week. Students were evenly distributed across three categories of grade point averages; below 3.0, between 3.00 and 3.35 and above 3.35. Fifty-eight percent were taking a junior level accounting course, while 42% were taking a senior level accounting course. A majority of the students responding, 82%, were enrolled in four or five courses for the semester. Finally, 64% of the students were from universities with very difficult entrance requirements. Any observation differences between feedback groups occur due to normal non-responses on various survey items.

TESTS OF HYPOTHESES

We used the results of our survey to develop data for the test of hypotheses. First, we developed a reasonable basis for comparing the responses for a number of students in each class with their respective instructor's responses. A mean average for each class was calculated for each student survey response item. For example, a mean average was calculated for each of the 18 class sections for the student question "how many hours did you study for this course?" or labeled *student reported study hours (SR)*. Each class SR mean average was then matched with the respective instructor's response for that class. In the first hypothesis, each

Table 2. Student Demographic Data

Demographic	Frequency	Percent
Sex		
Female	162	47%
Male	179	53
Total	341	100%
Age		
21 years or younger	248	74%
22 years or older	87	26
Total	335	100%
Marital Status		
Not Married	304	89%
Married	37	11
Total	341	100%
Ethnicity		
White	271	80%
Minority	68	20
Total	339	100%
Weekly Hours Worked		
Not gainfully employed	164	49%
Working 1–19 hours	96	28
Working 20 hours or more	78	23
Total	338	100%
Grade Point Average (GPA)		
2.99 and below	110	33%
3.00 to 3.35	114	35
3.36 and higher	106	32
Total	330	100%
Accounting Course Level		
Junior level	198	58%
Senior level	143	42
Total	341	100%
Course Load		
Four courses or less	76	22%
Five courses	204	60
Six courses or more	61	18
Total	341	100%
Entrance Requirements		
Very Difficult	219	64%
Moderately or Minimally	122	36
Total	341	100%

Note: Any observation differences between feedback groups occur due to normal nonresponses on various survey items.

class SR mean average was matched with the instructor response for IB, instructor's belief of study hours. A data set of 18 matched pairs resulted for the 18 class sections. Data sets were created in a like manner for other hypotheses based on overall responses by class section.

We used a paired t-test procedure to test for significant differences, or **gaps**, between study hour responses based on each hypothesis. Testing the differences in this manner is a conservative approach as it results in the fewest degrees of freedom in the statistical procedure compared to an independent t-test procedure. Paired t-tests are robust against violations of normality and homogeneity of variances (Huck et al. 1974, 197). Huck et al. (1974, 197) also considers paired t-tests more powerful than nonparametric tests unless there is evidence to suggest extreme nonnormality and variances that are heterogeneous. Based on a sample size of 18 class sections, we used a comparative nonparametric test procedure to verify the results of the paired t-test procedure. A Wilcoxon matched-pairs signed-rank test procedure provided similar results for all hypotheses.

The second phase in analyzing the hypotheses was to consider the study hour responses based on student demographics. This phase necessitated developing a second data set aside from the first data set described above. In this second data set we matched each student's response to her/his respective instructor's response. For example, an instructor's single response for her/his class was matched to each student response in that class section. By matching the responses in this manner we were able to obtain worthwhile information based on student demographics. This approach is reasonable since the instructor's response is based on all the students in his/her class, and we asked for the response without prejudice to any one demographic group. We used paired t-test procedures to test for significant differences in this second data set. The resulting sample size of 341 represents the number of students surveyed after eliminating students with unusual circumstances. Any observation differences between hypotheses and feedback groups occur due to normal nonresponses on various survey items.

Hypothesis 1

Table 3 presents the mean differences and levels of significance for the first hypothesis, H1, which tests the *study gap*. The first hypothesis seeks to determine if accounting students report they study (SR) more than accounting instructors believe they study (IB), that is, SR > IB. Overall, there was a significant positive mean difference of .9 hours per week ($t = 1.714$, $df = 17$, $p \leq .05$). Significant positive differences existed for all demographic variables except for male students, students with a GPA of 3.00 or 3.35, senior level students, and students at schools with moderate to minimum entrance requirements.

Table 3. Study Gap (H1): T-tests of Differences between Student's Reported (SR) Study Hours and Instructor's Belief (IB) of Study Hours

Demographic	N	SR Mean	IB Mean	Difference	t
Overall by Class Students	18	5.3	4.4	.9	1.714[*]
Sex					
Female	179	5.8	4.6	1.2	4.173[**]
Male	161	4.5	4.5	.0	.338
Age					
21 years or younger	248	4.9	4.5	.4	2.057[*]
22 years or older	87	6.2	4.6	1.6	3.293[**]
Marital Status					
Not Married	303	4.9	4.5	.4	2.123[*]
Married	37	7.8	5.1	2.7	3.999[**]
Ethnicity					
White	270	5.3	4.7	.6	2.785[**]
Minority	68	4.9	3.9	1.0	2.127[*]
Work Load					
Not gainfully employed	164	5.4	4.7	.7	2.652[**]
Working 1-19 hours	95	4.6	4.1	.5	1.578[a]
Working 20 hours or more	78	5.3	4.6	.7	1.473[b]
GPA					
2.99 and below	110	4.5	4.0	.5	1.699[*]
3.00 to 3.35	113	5.0	4.6	.4	1.198
3.36 and higher	106	5.8	4.9	.9	2.333[*]
Accounting Course Level					
Junior level	198	6.4	5.2	1.2	4.750[**]
Senior level	142	3.5	3.6	−.1	.507
Course Load					
Four courses or less	76	5.9	4.6	1.3	2.553[**]
Five courses	203	5.0	4.6	.4	1.778[*]
Six courses or more	61	5.0	4.3	.7	1.784[*]
Entrance Requirements					
Very Difficult	122	6.0	4.5	1.5	3.901[**]
Moderately and Minimally	218	4.7	4.5	.2	1.032

Notes: [a] $p = .06$;
[b] $p = .07$;
[*] $p \leq .05$;
[**] $p \leq .01$ (one-tailed)
Any observation differences between feedback groups occur due to normal non-responses on various survey items.

Hypothesis 2

The *study expectation gap* hypothesis, H2, is presented in Table 4. H2 seeks to determine if accounting students believe they should study (SB) more than accounting instructors expect them to study (IE), that is, SB > IE. Overall,

Table 4. Study Expectation Gap (H2): T-tests of Differences between Student's Belief (SB) and Instructor's Expectation (IE) by Study Hours

Demographic	N	SB Mean	IE Mean	Difference	T
Overall by Class Students	18	8.1	7.2	.9	1.686[a]
Sex					
Female	169	8.4	7.3	1.1	3.396**
Male	157	7.5	6.8	.7	1.998**
Age					
21 years or younger	235	7.8	7.0	.8	2.740**
22 years or older	86	8.5	7.2	1.3	2.583**
Marital Status					
Not Married	289	7.9	7.0	.9	3.423**
Married	37	8.7	7.5	1.2	1.803**
Ethnicity					
White	260	7.7	7.2	.5	2.070**
Minority	64	9.2	6.6	2.6	3.976**
Work Load					
Not gainfully employed	156	7.8	7.3	.5	1.602
Working 1-19 hours	94	7.8	6.6	1.2	2.678**
Working 20 hours or more	73	8.7	7.2	1.5	2.583**
GPA					
2.99 and below	103	8.4	6.6	1.8	3.871**
3.00 to 3.35	109	7.7	7.1	.6	1.446
3.36 and higher	104	7.8	7.6	.2	.669
Accounting Course Level					
Junior level	190	9.1	8.3	.8	2.434**
Senior level	136	6.4	5.4	1.0	3.164**
Course Load					
Four courses or less	73	8.3	7.4	.9	1.768*
Five courses	197	7.9	7.1	.8	2.805**
Six courses or more	56	7.9	6.8	1.1	1.927**
Entrance Requirements					
Very Difficult	115	9.1	7.3	1.8	3.845**
Moderately and Minimally	211	7.4	7.0	.4	1.604[a]

Notes: [a] $p = .06$;
      * $p \leq .05$;
      ** $p \leq .01$ (one-tailed)
      Any observation differences between feedback groups occur
      due to normal nonresponses on various survey items.

there was a significant positive mean difference of .9 hours per week ($t = 1.686$, $df = 17$, $p = .06$). Significant positive differences existed for all demographic variables except for students not gainfully employed, and students with a GPA of 3.00 or higher.

Table 5. Motivation Gap (H3): T-tests of Differences between Student's Belief (SB) and Student's Reported (SR) by Study Hours

Demographic	N	SB Mean	SR Mean	Difference	T
Overall by Class Students	18	8.1	5.3	2.8	7.166[**]
Overall	325	8.0	5.2	2.8	13.108[**]
Sex					
Female	169	8.4	5.9	2.5	8.128[**]
Male	156	7.6	4.6	3.0	10.764[**]
Age					
21 years or younger	235	7.8	4.9	2.9	13.042[**]
22 years or older	86	8.5	6.2	2.3	4.496[**]
Marital Status					
Not Married	288	7.9	4.9	3.0	14.032[**]
Married	37	8.7	7.7	1.0	1.263
Ethnicity					
White	259	7.7	5.3	2.4	11.008[**]
Minority	64	9.2	4.9	4.3	7.507[**]
Work Load					
Not gainfully employed	156	7.8	5.5	2.3	7.680[**]
Working 1-19 hours	93	7.9	4.7	3.2	9.675[**]
Working 20 hours or more	73	8.7	5.3	3.4	6.101[**]
GPA					
2.99 and below	103	8.4	4.6	3.8	11.083[**]
3.00 to 3.35	108	7.7	5.0	2.7	7.428[**]
3.36 and higher	104	7.8	5.9	1.9	4.796[**]
Accounting Course Level					
Junior level	190	9.1	6.5	2.6	8.989[**]
Senior level	135	6.4	3.4	3.0	9.917[**]
Course Load					
Four courses or less	73	8.3	5.9	2.4	4.553[**]
Five courses	196	8.0	5.0	3.0	11.278[**]
Six courses or more	56	7.8	5.1	2.7	5.777[**]
Entrance Requirements					
Very Difficult	115	9.0	6.1	2.9	6.617[**]
Moderately and Minimally	210	7.4	4.7	2.7	12.265[**]

Notes: [*] $p \leq .05$;
[**] $p \leq .01$ (one-tailed)
Any observation differences between feedback groups occur due to normal nonresponses on various survey items.

Hypothesis 3

The *motivation gap* hypothesis, H3, is presented in Table 5. H3 seeks to determine if accounting students believe they should study (SB) more than they report they study (SR), that is, SB > SR. Overall, there was a significant positive mean difference of 2.8 hours per week ($t = 7.166$, df $= 17$, p .01). Significant positive differences existed for all demographic variables with $p \leq .01$) the exception of married students that showed no significant difference.

Table 6. Performance Gap (H4): T-tests of Differences between Instructor's Expectation (IE) of Study Hours and Student's Reported (SR) Study Hours

Demographic	N	IE Mean	SR Mean	Difference	t
Overall by Class Students	18	7.2	5.3	1.9	4.177**
Sex					
Female	179	7.3	5.8	1.5	5.573**
Male	161	6.8	4.5	2.3	9.072**
Age					
21 years or younger	248	7.0	4.8	2.2	10.745**
22 years or older	87	7.3	6.2	1.1	2.364**
Marital Status					
Not Married	303	7.0	4.8	2.2	11.208**
Married	37	7.6	7.8	−.2	−.339
Ethnicity					
White	270	7.2	5.2	2.0	9.058**
Minority	68	6.6	4.9	1.7	4.130**
Work Load					
Not gainfully employed	164	7.2	5.4	1.8	6.442**
Working 1-19 hours	95	6.6	4.6	2.0	6.526**
Working 20 hours or more	78	7.3	5.3	2.0	4.485**
GPA					
2.99 and below	110	6.6	4.5	2.1	6.432**
3.00 to 3.35	113	7.1	5.0	2.1	6.761**
3.36 and higher	106	7.5	5.8	1.7	4.603**
Accounting Course Level					
Junior level	198	8.3	6.4	1.9	7.123**
Senior level	142	5.4	3.5	1.9	7.271**
Course Load					
Four courses or less	76	7.4	5.9	1.5	3.187**
Five courses	203	7.1	5.0	2.1	9.096**
Six courses or more	61	6.7	5.0	1.7	4.077**
Entrance Requirements					
Very Difficult	122	7.2	6.0	1.2	3.492**
Moderately and Minimally	218	7.0	4.7	2.3	10.434**

Notes: * $p \leq .05$;
** $p \leq .01$ (one-tailed)
Any observation differences between feedback groups occur due to normal nonresponses on various survey items.

Hypothesis 4

Table 6 presents the results of testing H4, which tests the *performance gap*. H4 seeks to determine if accounting instructors expect them to study (IE) more than students report they study (SR), that is, IE > SR. Overall, there was a significant positive mean difference of 1.9 hours per week ($t = 4.177$, $df = 17$, $p < .01$). Significant positive differences existed for all demographic variables with $p < .01$ except for married students whose difference was not significant.

Hypothesis 5

Table 7 presents the results of testing H5, which tests the *cynicism gap*. H5 seeks to determine if accounting instructors expect them to study (IE) more than accounting instructors believe they study (IB), that is, IE > IB. Overall, there was a significant positive mean difference of 2.8 hours per week based on instructors' responses ($t = 8.200$, $df = 17$, $p < .01$). Significant positive differences existed for all faculty demographic variables.

Table 7. Cynical Gap (H5): T-tests of Differences between Instructor's Expectation (IE) by Study Hours and Instructor's Belief (IB) by Study Hours

Demographic	Instructors	N Classes	IE Mean	IB Mean	Difference	t
Overall by Class	12	18	7.2	4.4	2.8	8.200**
Faculty Sex						
Female	4	6	5.8	3.8	2.0	4.472**
Male	8	12	7.9	4.7	3.2	7.479**
Faculty Rank						
Assistant Professor	5	7	5.6	3.3	2.3	4.824**
Associate & Full Professor	7	11	8.3	5.2	3.1	6.773**
Years Teaching Course						
6 years or less	6	8	7.1	3.9	3.2	6.619**
More than 6 years	6	10	7.3	4.9	2.4	5.308**
Years at University						
9 years or less	6	8	6.9	3.6	3.3	8.881**
More than 9 years	6	10	7.5	5.1	2.4	4.609**
Course Level						
Junior level	6	12	7.5	4.5	3.0	12.186**
Senior level	6	6	6.6	4.3	2.3	2.539*
Entrance Requirements						
Very Difficult	5	9	6.9	4.4	2.4	7.234**
Moderately and Minimally	7	9	7.5	4.4	3.1	5.292**

Notes: *$p \leq .05$;
** $p \leq .01$ (one-tailed)

Summary of First Five Hypotheses

While we found significant differences across all hypotheses tested, the greatest and most consistent differences occur when testing hypotheses three, four, and five for the *motivation, performance,* and *cynicism gaps* respectively. These three study gaps show that the differences for all but the 'married student' demographic characteristics for the *motivation and performance gaps* are significant between categories where SB > SR, IE > SR, and IE > IB. Differences ranged from 1.9 to 2.8 hours per week overall. The only demographic characteristic that was not different was married students where instructors' expectation (IE) and students' reported study hours (SR) were statistically the same.

The overall difference for hypothesis two, the *study expectation gap*, was .9 hours per week when comparing student's beliefs about study time (SB) with their instructor's expectation (IE). We found the largest differences for *study expectation gap*, ranging from 1.5 to 2.6 hours per week, for students gainfully employed (1.5), students with a GPA below 3.00 (1.8 hours), students enrolled in schools with very difficult entrance requirements (1.8 hours), and minority students (2.6 hours).

The overall difference, .9 hours per week, we found was for hypothesis one, the *study gap*, comparing students' reported hours (SR) to instructors' beliefs of student study time (IB). The largest differences for the *study gap*, ranging from 1.5 to 2.7 hours per week. We found them for students enrolled in schools with very difficult entrance requirements (1.5 hours), students 22 years old or older (1.6 hours), and married students (2.7 hours).

All five study gaps are represented in Figure 3. This parallel of Figure 2 provides an easy to understand and vivid description of the differences between students and professors beliefs and actions.

Finally, we conducted additional statistical tests to determine if the demographic variables could explain any patterns for these differences. These tests revealed very little, if any, explanation as to why these differences exist. These results point to the complexity of *study time gap* and the perceptions that both students and instructors have toward study time. They also suggest that more research is needed to isolate these differences in future studies.

ADDITIONAL HYPOTHESES AND RESULTS

The study gaps we noted do not reveal what stance instructors should take— "study more" or "study less." Should instructors, when they communicate to their students, inform them that they are expected to study long hours to motivate them? Will instructors be more successful motivating their students if the students are told to study less?

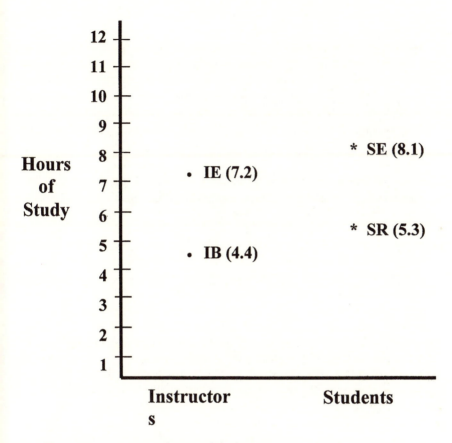

Key: IE is time instructors expect students to study for their course.
IB is time instructors believe student actually study for their course.
SE is the time students believe their instructors expect them to study for their course.
SR is the time students actually report studying for their course.

Figure 3.

To answer these questions we divided our data into two halves. The instructors who had greater than average expectations of their students were grouped into one pool ("high expectations") and instructors who had expectations below the average were grouped in the second pool ("low expectations"). We formulated the following hypotheses:

H6: Accounting students who have instructors with high expectations believe they should study more than students who have instructors with low expectations.

and:

H7:   Accounting students who have instructors with high expectations study
fewer hours than students who have instructors with low expectations.

The tests of these hypotheses in Table 8 reveal that students who have instructors with high expectations believe they should study more and report spending
more time studying more than students who have instructors with low expectations. Students who had instructors with high expectations studied 2.8 hours more
than did students who have instructors with low expectations. The difference
between the amount of time students believed they should study and the amount
of time they reported studying was also different if their instructors had high or
low expectations (SB-SR was 3.3 for students who had instructors with low
expectations and 2.6 for students who had instructors with high expectations).
The *motivation gap* lessens if instructors have high expectations.

IMPLICATIONS

The results support the *study time gap model*. The greatest difference of 2.8 hours
per week exists for the *motivation gap* between what time students believe they
should study for a course and the time they report studying (8.1 vs. 5.3 hours overall). The size of this gap suggests that students may place unrealistic expectations
on themselves to study, that they are not getting quality study time when they do
study, or they are not motivated by their instructor to study properly.

Another large difference of 2.8 hours per week exists between the instructors'
expectation and the instructors' belief of how long their students study (7.2 vs. 4.4

Table 8.   H6 and H7 T-tests of Differences between Instructor's Expectation
(IE) and Students' Belief (SB) and Students' Reported (SR) of Study Hours

Hypothesis H6	Classes/ Instructors	Students Assigned	SB Mean	IE Mean	Difference	t
Instructors with Low Expectations	9/5	174	7.0	5.4	1.6	5.358[**]
Instructors with High Expectations	9/7	152	9.1	9.0	.1	.236
Hypothesis H7						
Instructors with Low Expectations	9/5	184	5.4	3.7	1.7	8.151[**]
Instructors with High Expectations	9/7	156	9.0	6.9	2.1	6.411[**]

Notes: [*] $p \le .05$;
         [**] $p \le .01$ (one-tailed)

hours overall). This *cynicism gap* confirms the finding that instructors perceive that their students do not study enough and supports the *motivation gap* that students may not be motivated to study sufficiently. The *performance gap* shows a difference of 1.9 hours per week exists between the instructors' expectation and the students' reported study hours (7.2 vs. 5.3 hours overall). While students' study performance is below expectations, instructors seem to contribute to this *performance gap* since they make relatively good estimates of their students' study time per week. This result suggests that instructors need to think carefully about time-on-task when giving assignments if they believe they have good estimates of their students' study time. In addition, more difficult assignments might be divided into smaller tasks to help students with time management.

Students overestimated the time they thought they should study by almost a full hour over what their instructors believed was the necessary study time, referred to as the *study expectation gap* (8.1 vs. 7.2 hours overall). Instructors underestimated their students' study time, leading to a *study gap*. Instructors on average underestimated their students' study time by about an hour per week (5.3 vs. 4.4 hours overall).

CONCLUSION, LIMITATIONS, AND AREAS FOR FUTURE RESEARCH

Students do need to spend more time studying. Our results demonstrate that a *study time gap* exists. Students do not study as much as their instructors believe they should in order to master course material. However, students report they study more than their instructors believe they do. Students also overestimate how much they need to study. Students and instructors must communicate better so students can set realistic study goals and be motivated to study the appropriate amount of time to master the skills being taught to them.

The results of this study do not show how to motivate students to study more. Nor do they address the question of how some students study more effectively than others. There may be several possible explanations for students' sub-optimal study behavior. However, improved communications between instructors and their students should lead to better student study habits and eventual success.

The study was regional in focus. All of the classes surveyed were in universities located in the southeastern part of the United States. There is no reason why these results should not apply on a national basis, but a replication of a broader survey population would be interesting to determine that with certainty.

We chose to request responses after the fact, and we surveyed students only once during the semester. A subsequent study could repeat the survey many times during the semester and include exam weeks. Researchers could ask students to keep a log and see if results differ from what we found.

We find it hard to believe that instructors tell students how much time they need to study outside of class unless the instructors believe they are giving students good advice. Then why is there a "cynicism gap?" This is a question for future research.

Future research can progress in several directions. More demographic data could be collected in controlled settings to identify the characteristics of students who correctly match their study time to the required effort for mastering course work. Once those characteristics are identified future research can attempt to discover ways to motivate all students to match their efforts with the optimal time requirements. Investigations into the optimal study behavior also would be very interesting, but are probably well beyond the scope of this type of survey research.

A second area of research could pursue the ties between motivation and behavior. If instructors communicate less rigorous time goals to their students, will student study time increase? Will student goals shift downward and study time decrease? Also, do some instructors motivate their students better than other instructors? Findings would allow more instructors to adopt strategies, which might result in a better match in student study time and the optimal study time determined by their instructors. Researchers also could examine the "why" questions. We have shown that gaps exist, so the next step could be to ask why.

APPENDIX A

Student Study Practices Survey

We are conducting a statewide survey of student study practices. Please answer the following questions as best you can. We will be happy with your best estimates.

Course number for this course _____ Date _____

LAST WEEK (and please consider **ALL** seven days) how many hours did you study for this course? _____

Was there anything unusual last week that would have caused you to study more or less for this course? (please explain) _____

How many hours do you think your instructor expected you to study? _____

DEMOGRAPHIC INFORMATION

1. Sex (check one) Male _____ Female _____ 2. Your age _____

3. Marital Status (check one)   4. Year in school (check one)
   Married _____ Not Married _____   Junior _____ Senior _____

5. Overall GPA _____

6. Number of hours per week you work for pay (wages, salary etc.) _____

7. Number of credit hours in which you are currently enrolled _____

8. Number of courses in which you are currently enrolled _____

9. Ethnic background (check one)
   _____ White (included Arabian)
   _____ Black (includes Jamaican, Bahamian, and other Caribbeans of
   African but not Hispanic or Arabian decent)
   _____ Hispanic (includes persons of Mexican, Puerto Rican, Central or
   South American or Spanish Origin)
   _____ Asian (includes Pakistanis, Indian and Pacific Islanders)
   _____ Native American Indians (includes Eskimos)

APPENDIX B

Faculty Portion Student Study Practices Survey

We are conducting a statewide survey of student study practices. Please answer the following questions as best you can. We will be happy with your best estimates.

Course number for this course _____ Date _____

LAST WEEK (and please consider **ALL** seven days) how many hours do you think your students (on the average) studied for this course? _____

Was there anything unusual last week that would have caused them to study more or less for this course? (please explain)_____

How many hours did you expect them to study **LAST WEEK**? _____

DEMOGRAPHIC INFORMATION

1. Sex (check one) Male ____ Female ____

2. Your faculty rank (check one) Assistant ____ Associate ____ Full ____

3. Number of years you have taught this course ____

4. Number of years at this University ____

5. Ethnic background (check one)
    ____ White (included Arabian)
    ____ Black (includes Jamaican, Bahamian, and other Caribbeans of
    African but not Hispanic or Arabian decent)
    ____ Hispanic (includes persons of Mexican, Puerto Rican, Central or
    South American or Spanish Origin)
    ____ Asian (includes Pakistanis, Indian and Pacific Islanders)
    ____ Native American Indians (includes Eskimo)

ACKNOWLEDGMENT

We would like to acknowledge the review comments of Mohamed Bayou (University of Michigan—Dearborn), Sue Ravenscroft (Iowa State), James McMillan (Virginia Commonwealth University) and several anonymous reviewers at the Mid-Atlantic Region and AAA Annual Meetings.

NOTES

1. In our study we use "hours studying" as a surrogate for "effort." This substitution addresses the quantity of effort not the quality of the effort.

2. Unusual circumstances were those circumstances over which the student had little to no control over during the week of reported study time. Of the 22 student surveys eliminated, nine were eliminated due to sickness or illness, eight due to students being out-of-town during the week, and five due to a personal crisis such as death or extra work from a new job.

REFERENCES

Adair, J.G. et al. 1987. *Hawthorne Control Procedures in Educational Experiments: A Reconsideration of their Use and Effectiveness.* Annual Meeting of the American Educational Research Association (Washington, DC, April 20-24, 1987).

Britton, B.K., and A. Tesser. 1991. Effects of time management practices on college grades. *Journal of Educational Psychology* 83(3): 405-410.

Frisbee, W.R. 1984. Grades and academic performance by university students: A two-stage least squares analysis. *Research in Higher Education* 20(3): 345-365.

Gleason, J.P., and W.B. Walstad. 1988. An empirical test of an inventory model of student study time. *Journal of Economic Education* 19(4): 315-321.

Huck, S.W., W.H. Cormier, and W.G. Bounds, Jr. 1974. *Reading Statistics and Research*. New York: Harper Collins.

Ibrahim, M.E. 1989. Effort-expectation and academic performance in managerial cost accounting. *Journal of Accounting Education* 7: 57-68.

Kelly, D.K. 1992. *Part-Time and Evening Faculty: Promoting Teaching Excellence for Adult Evening College Students*. Fullerton College Office of Instruction, Fullerton, CA.

Kember, D. et al. 1996. An examination of the interrelationships between workload, study time, learning approaches, and academic outcomes. *Studies in Higher Education* 21(3): 347-358.

Lawler, E.E. 1967. Antecedent attitudes of effective managerial performance. *Organizational Behavior and Human Performance* 2: 122-142.

Ott, R.L. 1988. Pretest reviews in intermediate accounting. *Issues in Accounting Education* 3(2): 378-387.

Ott, R.L., D.S. Deines, and D.P. Donnelly. 1988. The use of a fundamental practice set in intermediate accounting. *Issues in Accounting Education* 3(1): 131-138.

Parry, R.W. 1990. The impact of assigned study groups on study effort and examination performance. *Issues in Accounting Education* 5(2): 222-239.

Peterson's 1999 Guide to Four Year Colleges. Princeton, NJ: Peterson's.

Reed, P. et al. 1984. The validity of an academic study time questionnaire for college students. *Educational & Psychological Measurement* 44(4): 1031-1036.

Thombs, D.L. 1995. Problem behavior and academic achievement among first-semester college freshmen. *Journal of College Student Development* 36(3): 280-288.

INTEGRATING RESEARCH INTO THE INITIAL AUDITING COURSE

Paul M. Clikeman

ABSTRACT

This article provides suggestions to integrate research into the initial auditing course. The benefits of integrating teaching and research include making presentations more stimulating, updating course content, helping influence accounting practice, and improving students' critical thinking skills. The article describes research studies relevant to topics typically covered in an initial auditing course. Suggestions for classroom exercises and homework assignments, based on audit research, appear throughout the article.

Advances in Accounting Education, Volume 2, pages 113-130.
Copyright © 2000 by JAI Press Inc.
All rights of reproduction in any form reserved.
ISBN: 0-7623-0515-0

INTRODUCTION

One of the major consequences of introducing research is to alter students' perception of accounting as a discipline. Accounting does involve judgments and research has been conducted that provides evidence on the nature and consequences of those judgments. Other fields, such as finance, incorporate nontechnical summaries of research into their introductory (and more advanced) texts to a much greater extent than we appear to in accounting. The reason for this disparity is not obvious (Beaver 1984, 37).

This article describes how I integrate audit research into an initial auditing course. Through lectures and assigned readings, I expose students to 10 or 12 audit research studies during a semester. The research studies help students appreciate the complexity of auditing judgments and understand how economic and legal pressures might influence auditors' behavior.

The initial auditing course I teach is required for accounting majors. Students usually take it during their senior year and it is the only auditing course most of the students will take. The course focuses on financial statement audits. Major topics include generally accepted auditing standards, audit reports, the AICPA *Code of Professional Conduct*, auditors' legal liability, risk and materiality, internal control evaluation, and audit evidence. Many textbooks are available that cover these topics and prepare students to pass the CPA exam, but I felt the books did not describe adequately the complexity of audit judgments or the many environmental variables that can influence auditors' decisions. Approximately half of my students will become auditors after they graduate. I wanted these future practitioners to see the results of past audits, understand how audit practice has changed in recent years, and be prepared for the economic and legal pressures they will face as auditors.

Integrating research studies into the auditing course has helped me describe the rich environment in which auditing is practiced. Surveys of financial statement users illustrate the expectation gap that exists between auditors and members of the general public. Studies of audit work papers help illustrate how inherent risk varies across accounts and classes of transactions. Studies that investigate the reliability of audit evidence demonstrate the risks and tradeoffs involved in choosing audit procedures. Surveys of auditors describe the economic and legal pressures they face.

This article provides suggestions for integrating research into the initial auditing course. The benefits of integrating teaching and research include making presentations more stimulating, updating course content, helping influence accounting practice, and improving students' critical thinking skills. I describe audit research studies relevant to four topics typically covered in an initial auditing course. Suggestions for classroom exercises and homework assignments, based on audit research, appear throughout the article.

BENEFITS OF INTEGRATING RESEARCH

Integrating research and teaching is the norm in most academic disciplines. Review of almost any introductory psychology, physics, biology, or organizational behavior textbook will find descriptions of and references to relevant research. For 25 years, accounting educators have called for similar integration of accounting research into accounting education (Sterling 1973; Beaver 1984; Wright 1997). The benefits of integrating accounting research into the classroom appear below.

Burilovich (1993, 309-310) describes how synthesizing research and teaching provides a more "dynamic and intellectually stimulating" presentation of material that otherwise might be mundane. Archival studies that describe actual audit results provide interesting examples of the misstatements auditors discovered (Houghton and Fogarty 1991; Loebbecke et al. 1989). Behavioral studies of auditors' judgments demonstrate the types of decisions auditors make, and the environmental pressures that might influence their judgments (Pratt and Stice 1994; Bell and Wright 1997).

Integrating teaching and research also can help instructors update their course content (Dopuch 1989, 1). The competitive and legal environment of auditing has changed dramatically in recent years. After enjoying substantial growth and profitability during the 1970s and 1980s, accounting firms have experienced stagnation in inflation-adjusted audit revenue during the 1990s (Berton 1996, B1). Accountants are reducing audit fees to win and retain clients and are becoming increasingly dependent on consulting services to earn profits (Public Accounting Report 1991, 1). Expenditures for settling and defending lawsuits now consume more than 10% of auditing and accounting revenues in the United States (Pany and Whittington 1997, 135). The results of research studies can help students understand how increased competition and the threat of litigation might influence auditors' judgments.

Introducing research into the classroom also may help researchers influence future accounting practice. One reason accounting research has had limited impact on accounting practice is that many practitioners are ill prepared to read accounting research (Leisenring and Johnson 1994, 75-76). Exposing students to audit research may make future practitioners more receptive to accounting researchers' conclusions and recommendations.

Finally, discussing research in the classroom may help students improve their critical thinking skills. Markham (1991, 466-467) discusses the importance of exposing students to research.

> As part of a liberal education, students need to understand the logic of scientific research. They need to learn how to formulate generalizations about relationships between variables based on careful data gathering, rather than on overgeneralization from personal experience or on blind acceptance of cultural myths.

Table 1. Research Studies Relevant to Initial Auditing Course

Topic	Research Studies	Major Findings	Lessons for Students	Classroom Exercises/ Homework Assignments
Expectation Gap	Lowe and Pany (1993)	Members of general public expect auditors to search for fraud, act as public watchdog, and insure investors against losses.	Documents the problem of the expectation gap. Valuable for introducing SAS 82.	Assign students to complete the questionnaire and compare their answers with professional standards and auditors' answers.
	Loebbecke, Eining, and Willingham (1989)	Summarizes 354 material frauds reported by KPMG partners.	Identifies most common types of fraud and accounts most likely to be misstated. Describes red flags associated with fraud.	Assign students to read about a recent fraud and identify red flags associated with the fraud.
Risk Analysis	Houghton and Fogarty (1991)	Summarizes all misstatements detected during 480 DH&S audits.	Helps students see how inherent risk varies across accounts and classes of transactions.	Assign students to read an annual report and identify accounts and transactions with highest risk of misstatement.
	Caster (1992)	Responses to accounts receivable confirmations were incomplete and biased.	Auditors should maintain skepticism when evaluating confirmations. Includes suggestions for improving reliability of confirmations.	
Reliability of Audit Evidence	Pany and Wheeler (1992)	Simple analytical review techniques were relatively ineffective compared with ratio analysis and statistical techniques.	Demonstrates the tradeoff between efficiency and effectiveness in selecting auditing procedures.	Assign students to perform analytical procedures to detect seeded misstatements.
	Pratt and Stice (1994)	Auditors, in an experiment, recommended more extensive testing and higher audit fees when litigation risk was assessed as high.	Valuable for discussing the importance of investigating a prospective client before accepting an engagement.	Assign students to investigate a public company and assess the auditor's engagement risk and litigation risk.
Auditor Litigation	Dalton, Hill, and Ramsay (1994)	Many former auditors report that litigation influenced their decision to leave public accounting.	Demonstrates how fear of litigation may influence the conduct of an audit. Valuable for discussing auditor liability and the effects of litigation on individual auditors.	

116

Learning to evaluate evidence critically and draw appropriate conclusions is an essential skill for auditing students to learn (AICPA 1997, AU Section 150.02). Oliverio (1984, 53-54) describes the similarities between auditing and scientific research. Auditors gather evidence to test financial statements just as researchers gather evidence to test hypotheses. Professors can use research studies to demonstrate the process of designing tests and examining evidence. Students can evaluate research studies to determine whether the evidence justifies the conclusions of each study.

EXAMPLES OF INTEGRATING RESEARCH

This section describes examples of integrating research into an initial auditing course. Table 1 lists research studies that examine: (1) the expectation gap, (2) risk analysis, (3) the reliability of audit evidence, and (4) auditor litigation. These four topics are among the most important concepts covered in the initial auditing course (Bryan and Smith 1997, 5). Brief descriptions of the research studies and suggestions for classroom exercises and homework assignments appear below.

The Expectation Gap

The *expectation gap* refers to the problem of the public expecting more and different services than auditors are willing to perform. Lowe and Pany [LP] (1993) document an expectation gap between accountants and members of the general public. One hundred and forty-one members of a county juror pool and 78 auditors answered eight questions regarding auditors' responsibilities and performance. The survey responses revealed substantial differences of opinion between jurors and auditors. Jurors believed much more strongly than auditors that the auditor's role includes actively searching for fraud, acting as a public watchdog, and insuring investors against losses. Lowe and Pany (1993, 59) concluded, "Jurors systematically expect more from auditors than auditors believe they provide."

For the last several semesters, I have used the LP study on the first day of class. I have the students read a copy of a standard, unqualified audit report and give them a few minutes to answer the eight questions in the LP study. For each question, I ask several students to share their response until I find two students with differing opinions. Then I ask the students to explain their responses. Other students volunteer or are asked to explain why they agree or disagree with the first two students' views. Question 4 (the auditor's role as public watchdog) and question 5 (the auditor's responsibility for fraud) usually generates the most active discussion as my students have widely varying opinions about the auditor's role.

As we discuss each question, I write the mean scores from the LP study on the blackboard. Usually, about two-thirds of my students give answers similar to the CPA responses. These students often are surprised to learn that the members of the juror pool expect so much from auditors in terms of finding fraud and protecting shareholders from losses. The remaining one-third of the students give answers more similar to the jurors' responses. These students view auditors as "watchdogs" and are surprised to learn that CPAs have a different opinion about their role. Discussing the LP study helps my students see on the very first day that auditing is a challenging profession practiced in a complex environment.

Risk Analysis

I spend a lot of time discussing risk analysis in my auditing course. Auditors evaluate risk so they can focus audit tests on areas that are most likely to be misstated. I want my students to be able to assess the risk of fraudulent financial reporting and identify accounts or classes of transactions that are most likely to be misstated. Research studies that document the nature and frequency of financial statement misstatements can help auditing students learn to assess audit risks.

In 1997, the AICPA issued SAS No. 82, "Consideration of Fraud in a Financial Statement Audit." The standard requires auditors to assess fraud risk on every audit engagement by considering the presence of risk factors (*red flags*) related to fraudulent financial reporting and misappropriations of assets. When introducing students to fraud risk, I always describe briefly the Loebbecke et al. [LEW] (1989) study. LEW surveyed KPMG Peat Marwick audit partners to learn about the frequency and characteristics of financial statement fraud. Their study summarizes 354 instances of fraud the KPMG partners reported and classifies the misstatements by type, client industry, and account. LEW Table 9 describes the risk factors that were present most frequently in the frauds the KPMG partners reported. The AICPA subsequently included many of these risk factors in SAS 82. Discussing the LEW study helps students see that the guidance offered in SAS 82 is based on or at least consistent with academic research.

The LEW paper describes a conceptual model to help auditors assess fraud risk. According to the model, the likelihood of fraud is a function of (1) management's motive to commit fraud, (2) the ethical character of management, and (3) conditions that allow opportunities to commit fraud. The model helps auditing students apply SAS No. 82 by describing how various red flags indicate motives for fraud, questionable management integrity, and opportunities for fraud.

One assignment I have used successfully is having students read a description, or watch a video, of a well-publicized fraud (e.g., ZZZZ Best, Phar-Mor, Crazy Eddie) and write a memorandum identifying the red flags that should have alerted

Table 2. Sample Homework Assignment Risk Analysis

Read "Auditors' Experience with Material Irregularities: Frequency, Nature, and Detectability," J.K. Loebbecke, M.M. Eining, and J.J. Willingham, *Auditing: A Journal of Practice and Theory 9* (Spring, 1989). Pay particular attention to the risk assessment model described in Figure 1 and the fraud indicators listed in Figure 2.

Read "Battle of the Books," Lee Berton, *The Wall Street Journal*, January 24, 1989, and "How Don Sheelen Made a Mess That Regina Couldn't Clean Up," John A. Byrne, *Business Week*, February 12, 1990. Identify the fraud indicators (*red flags*) that should have warned the auditors that Regina was a high risk audit.

Required:
Imagine you are the senior accountant in charge of the 1988 audit of Regina. As part of the audit planning process, the partner has asked you to prepare a written memorandum addressing the following two items.

- Assess the risk of fraudulent reporting at Regina. Use the Loebbecke/Eining/Willingham risk assessment model to organize your discussion of fraud risk factors.
- Identify the two or three accounts that have the highest risk of being materially misstated. Document clearly the factors that led you to conclude these accounts have high inherent risk.

the auditors to the possibility of fraud. Table 2 presents an example of one such homework assignment. Students enjoy reading about the frauds and usually do a good job of identifying fraud risk factors and high-risk accounts. I use this assignment to help students improve their written communication skills. Most of the grade is based on how clearly and concisely the students support their risk assessments. Students must rewrite and resubmit deficient memorandums just as auditors must satisfy the supervising partner's review notes before placing a memorandum in the audit work papers.

SAS No. 47, "Audit Risk and Materiality in Conducting an Audit," requires auditors to evaluate risk when determining the nature, timing, and extent of audit procedures. Identifying high-risk accounts and transactions can be a difficult task for auditing students who have little first-hand knowledge of financial misstatements. I use Houghton and Fogarty's [HF] (1991) research study to demonstrate to students how inherent risk varies across accounts and transactions. HF summarizes the misstatements detected during 480 Deloitte Haskins & Sells audits and classifies the misstatements by account, cause, and environmental characteristics. The primary finding is that non-systematically processed transactions, such as the acquisition or sale of an affiliate, have a disproportionately higher likelihood of error than systematically processed transactions. Auditors most likely will detect material misstatements in receivables, inventories, and accrued liability accounts.

Because the HF study is rather long, I usually describe it during a lecture rather than require students to read it on their own. I describe the research methodology and use one or two slides of tables to describe the study's major findings. I can summarize the study and its major findings in less than ten minutes.

HF (1989, 18) describes how Deloitte Haskins & Sells assigned an international task force to incorporate the study's findings into a revised audit approach. I read aloud selected paragraphs from page 18 to the students to demonstrate how accounting research can have a direct and important influence on audit practice.

I have used one homework exercise based on the HF study. I have students read a company's annual report and write a memorandum identifying the accounts and transactions with the highest inherent risk. Students must investigate a company's business operations and identify complex transactions. They also must identify accounts that require significant management estimates or involve complicated accounting issues. Students develop library research skills and learn to use electronic databases when they investigate the company and its industry. They also practice written communication skills when writing the memorandum. Often I have the students work in small groups to help them develop teamwork skills and to reduce the number of assignments I have to grade. Allowing students to work in groups usually improves the quality of the memorandums because they can help each other find the necessary information, and edit each other's work (Gersten 1995, 71).

The Reliability of Audit Evidence

After discussing audit risk, I begin teaching my students how to perform audit tests and gather audit evidence to test financial statement assertions. I find the following two research studies valuable in helping my students better understand the reliability of audit evidence and the effectiveness of audit tests.

Caster (1992) conducted a field study to test the reliability of accounts receivable confirmations. He mailed positive accounts receivable confirmation requests to a steel warehousing operation's customers but incorporated large and small overstatement and understatement errors into the account balances on the confirmation requests. Casters' results raise serious doubts about the reliability of accounts receivable confirmations. Customers reported only 47.2% of the errors in the receivables account balances to the auditors. The evidence the confirmations provided also was biased. Customers were more likely to report overstatement errors than understatement errors. Caster (1992, 74) concluded, "Practitioners must maintain professional skepticism when evaluating confirmation evidence to avoid over reliance."

SAS No. 67, "The Confirmation Process," requires auditors to evaluate the reliability of confirmations. I use the Caster study to demonstrate the limitations of accounts receivable confirmations. Because of response bias, confirmations provide only limited evidence about the valuation of accounts receivable.

The Caster study was written for a practitioner audience and is only four pages long. I have the students read it on their own before coming to class and have found that undergraduate accounting students can read the article easily. They have little difficulty comprehending the study and its conclusions. Classroom discussion is brief and focuses on situations where auditors would expect accounts receivable confirmations to be unreliable. We also discuss briefly the article's suggestions about how to write and mail confirmation requests to maximize the probability of receiving a response and to increase the reliability of the information received.

Pany and Wheeler [PW] (1992) investigated the effectiveness of common analytical review procedures. PW took quarterly financial statements from five single-industry companies and altered the account balances to reflect seven misstatements such as fictitious credit sales and depreciation expense miscalculations. PW then determined whether simple analytical procedure investigation rules, such as investigating a 10% change in an account balance or a 10% change in an appropriate ratio, would detect the misstatements. The results raise serious doubts about the effectiveness of simple analytical procedures. A "10% change" investigation rule only found 19% of the misstatements. A "5% change" investigation rule found more of the misstatements, but also led to the investigation of many accounts that were not misstated. PW (1992, 67) concluded, "...the auditor who uses simple percentage-change methods should place extremely limited reliance on their results."

I describe the PW study during a lecture and use it to show my students the tradeoff between effectiveness and efficiency that auditors must make when performing audit procedures. Auditors who use a low investigation threshold will detect more misstatements (i.e., be more effective), but also will investigate more accounts that are not misstated (i.e., be less efficient). The PW study helps demonstrate the importance of developing more sophisticated analytical procedures rather than simply relying on mechanical cutoff rules to investigate changes in account balances. PW found that ratio analysis and statistical techniques for identifying unusual fluctuations outperformed the simple account balance change analysis.

As a homework assignment, I have students select a company and enter two years of income statement and balance sheet information into a spreadsheet. The students then alter the current year financial statements to reflect various types of misstatements. For each misstatement, the students calculate the effect on the relevant ratios and determine whether common investigation thresholds would trigger investigation of the misstated accounts. This assignment helps auditing students visualize how various errors affect the financial statements. It also helps students begin to understand the effectiveness of simple analytical procedures.

Auditor Litigation

In 1992, the Big Six accounting firms issued a statement of position claiming that an "Epidemic of Litigation" threatened the existence of accounting firms of all sizes (Andersen et al. 1992, 1). Articles describing lawsuits against public accounting firms continue to appear frequently in popular business periodicals. Many of my students express concern about how these lawsuits are affecting the accounting profession. I use the following two research studies to discuss litigation's effect on auditors' behavior.

Pratt and Stice [PS] (1994) investigated whether auditors increase their audit fees and collect more audit evidence when they perceive litigation risk to be high. Two hundred and forty-three managers and partners from four of the then Big Six firms read descriptions of a fictitious prospective client's operations and financial condition, assessed the client's litigation risk, and recommended certain elements of an audit plan and a client fee. The participants assessed the litigation risk highest when the client's financial condition was poor. The auditors also recommended more extensive testing and higher audit fees when litigation risk was assessed as high.

I use the PS study to discuss the importance of investigating a prospective client before accepting an engagement. There is a positive relationship between client bankruptcies and lawsuits against auditors (Palmrose 1997, 67), and auditors are well advised to consider the client's financial condition before beginning the audit. I also use the PS study to reinforce the concept of audit risk. In certain situations, auditors may wish to gather more evidence than the minimum required by generally accepted auditing standards in order to reduce the likelihood of issuing an incorrect opinion.

Table 3 contains an example of a homework assignment requiring students to read the PS article and discuss its findings and practice implications. Even though the article was published in a top-tier academic journal, students do not need training in research design or statistical analysis to complete the assignment. The first three questions simply require students to identify the research question and

Table 3. Sample Homework Assignment Summary and
Analysis of a Research Study

Read "The Effects of Client Characteristics on Auditor Litigation Risk Judgments, Required Audit Evidence, and Recommended Audit Fees," J. Pratt and J.D. Stice, *The Accounting Review* 69 (October, 1994): 639-656.

Required:
Write a one- to two-page paper addressing the following questions.

1. What research question is the study designed to answer?
2. What variables were found to influence the auditor's assessment of litigation risk?
3. How did the subjects in the study adjust the audit plan when litigation risk was assessed as high? Is this behavior consistent with generally accepted auditing standards?
4. In 1992, the Big Six accounting firms claimed that an "epidemic of litigation" threatened the existence of accounting firms of all sizes. Using the results of the Pratt & Stice study, how would you expect auditors to react to the "epidemic"? How would clients be affected by the "epidemic"?
5. In 1995, Congress passed the Private Securities Litigation Reform Act. How would you expect auditors to react to the passage of this Act? How might clients benefit from the legislation? Is there any potential downside to the expected change in auditors' behavior?

major findings. The last two questions require students to exercise critical thinking skills by applying the research findings to recent events.

Dalton et al. [DHR] (1994) conducted a study to determine whether litigation is affecting auditors' decisions to leave public accounting. The researchers mailed questionnaires to 904 former managers and partners who had resigned recently from Big Six firms. Of the 211 respondents, approximately 29% reported that the threat of legal liability played a role in their decisions to leave their firms. More than 40% of the respondents reported that depressed profits because of litigation costs affected their decisions. Many respondents also reported that litigation risks are causing competent people to leave public accounting and that the threat of litigation is leading to over auditing.

The DHR study is easy for undergraduate students to read and understand. The study generates lively classroom discussion about auditor liability and the effects of litigation on individual auditors. Many of my students are considering careers in public accounting, and they enjoy discussing whether the litigious environment is making it difficult for public accounting firms to attract and retain qualified auditors.

GUIDELINES FOR INTEGRATING RESEARCH

Professors must integrate research carefully and properly to achieve the benefits described in the first section. There are several potential obstacles to overcome. First, many undergraduate accounting students have little formal training in scientific research and the scientific method. Such students are not familiar with the process of gathering evidence and drawing conclusions. Second, many accounting students do not have strong statistical skills. Such students may have difficulty understanding the methodology and analysis sections of some research studies. Finally, some students may question the relevance of auditing research or may resist reading articles or completing assignments that they perceive as "extra work" beyond the textbook material.

Brems (1994, 241-242) describes her "gentle approach" to introducing undergraduate students to psychology research. She begins by describing research studies during lectures. This first step familiarizes students with the topics that are being investigated within the discipline and introduces students to the research process. Later, Brems has students read research articles and identify each study's research questions and major conclusions. Students become familiar with the language and format of research articles but do not have to tackle the potentially confusing methodology and analysis sections. Finally, Brems' students read research articles and critique the data-collection procedures and the conclusions drawn from the data.

My own approach to integrating audit research is similar to Brems' approach. During a semester, I describe between eight and ten research studies during lectures. I concentrate on each study's research question and major findings and give intuitive explanations of the methodology. I can describe the research question and major conclusions of most studies in between five and ten minutes. The total class time devoted to auditing research is modest, yet I expose the students to many examples of audit research. My exams always include questions about the research studies discussed in class. Some less-motivated students will not pay attention to the class discussions unless they know the material will be covered on exams. Students generally perform as well on the questions about research as they perform on questions about other course material.

During a semester, I also have students read two or three research articles. I usually select relatively short articles published in practitioner journals (e.g., Lowe and Pany 1993; Dalton et al. 1994). Because I limit the reading assignments to a small number of short articles, I have not received significant student complaints about the quantity or difficulty of their workload. The practitioner articles usually employ relatively simple research methodologies and give good intuitive descriptions of the analysis. I have found that undergraduate students can read articles in practitioner journals on their own and understand the main conclusions. Having students read the articles outside of class conserves class time. I can lead the students in a discussion of the article's findings and practice implications without spending time describing the study. Grading class participation or holding students responsible for each article's contents on exams and quizzes helps ensure that students read them.

Having students read articles in practitioner journals (e.g., *Journal of Accountancy, The CPA Journal, Internal Auditor*) introduces students to the resources that will serve as their "textbooks" throughout their professional careers. Professors can prepare students for lifelong learning by helping students read and comprehend research in practitioner journals. Permission to copy articles may be obtained by contacting the publisher or the Copyright Clearance Center. Joseph et al. (1996, 79-80) describe guidelines for copying and distributing copyrighted material.

During a semester, I also assign one or two of the homework assignments described in Table 1. The two risk analysis assignments require students to investigate a company and prepare a written memorandum. These assignments help students apply research results and practice their writing skills. The Pany and Wheeler (1992) assignment requires students to perform simple analytical review procedures and practice their spreadsheet skills.

Professors who want to emphasize critical thinking skills might have students read an audit research study and prepare a written analysis of the study's methodology and conclusions. Girden (1996) describes how to read research articles with skepticism. She describes several common research mistakes including selection bias, experimenter expectancy, and confusing correlation with causation. Professors might describe these research flaws and have students critique an audit research study. Understanding common research mistakes may help students become better auditors. Auditors, like researchers, must learn to select representative samples when selecting transactions or items for testing. Auditors also must avoid allowing confirmatory bias to affect their evidence search (Bailey 1986, 27).

Beaver (1984, 36-37) offers sound advice for incorporating accounting research into the classroom. First, Beaver recommends that research should not be discussed in isolation; the research must be used in connection with materials that illustrate its relevance. My own students are interested primarily in material that will help them pass the CPA exam or that they think will help them in their careers. They are skeptical of material that is not covered in the textbook or that appears to be too theoretical. I am careful to choose research studies that relate directly to topics covered in the textbook (e.g., audit risk, analytical procedures), and I always discuss each study's practice implications. Second, Beaver recommends that professors stress the study's findings and implications rather than the research methods. As mentioned above, many accounting students do not have strong statistical skills. Spending too much time discussing the statistical analysis will cause some students to lose interest in the research or become lost in the methodological details. Although occasionally I lead the students in a discussion of a study's methodology as an exercise in evidence gathering and evaluation, I usually focus on each study's research question and major findings. Finally, Beaver recommends that professors remind students that most research findings are tentative. Discussing the tentative nature of research findings illustrates the concept of audit risk. Even after conducting a research study (audit) and evaluating the evidence, a researcher (auditor) might draw an incorrect conclusion. A professor might even assign students to find related research studies and determine whether they support or contradict the findings discussed in class.

SUMMARY AND CONCLUSIONS

Several commentators have recommended introducing more research into accounting education (Beaver 1984; Burilovich 1993; Dopuch 1989). The benefits of integrating teaching and research include making presentations more stimulating, updating course content, improving students' critical thinking skills, and providing students with insights into current accounting practice.

My own experience with integrating research into the initial auditing course has been positive. Most students are used to seeing research in their psychology, sociology, organizational behavior, marketing and other business courses, and accept readily that an auditing course includes examples of auditing research. Although they have seen examples of research in other business disciplines, most students say that my course is their first exposure to accounting research. For the first time, students see that research can help us understand accountants' behavior better. Students discover that accounting is a legitimate academic discipline and learn that accountants make many judgments beyond just learning and following generally accepted standards.

In my experience, the primary benefits of integrating audit research have been helping students understand the results of past audits, the effectiveness of various audit procedures, and the economic, legal, and social factors that might influence auditor judgments. Integrating research into the auditing course has helped me change the focus of the course from "Here's the material you must memorize to pass the CPA exam" toward "Let's examine and discuss the judgments auditors make while performing an audit." My interest in teaching the course has increased tremendously. Students seem to welcome our periodic 10-minute discussions of audit research as nice breaks from lectures about professional standards, internal control procedures, and audit tests.

This article describes four examples of integrating audit research into an initial auditing course. The research studies examine the expectation gap, risk analysis, the reliability of audit evidence, and auditor litigation. The body of audit research is quite extensive, and many other opportunities exist for integrating research into the auditing course. The Appendix describes nine more research studies that may be relevant to the initial auditing course. Usually, I discuss these studies during lectures, but the Appendix describes possible exercises for several of the studies. Bell and Wright (1995) provide a comprehensive review of audit research and its impact on audit practice. Interested instructors can find a wealth of research studies to supplement the initial auditing course.

APPENDIX

Additional Research Studies Applicable to Initial Auditing Course

Topic	Research Studies	Major Findings	Classroom Exercises/ Homework Assignments
Audit Reports	Geiger (1994)	A survey of 178 bank loan officers found low satisfaction with the wording of the auditor's report.	Assign students to answer the 11 questions in the Hermanson et al. survey and compare their answers with the answers given by members of the American Association of Individual Investors.
	Hermanson, Duncan, and Carcello. (1991)	A survey of stockholders found that many did not understand the information in the standard auditor's report. Article includes suggestions for improving the audit report's wording.	
Audit Litigation	Palmrose (1997)	The 20 largest U.S. accounting firms have been sued more than 1,000 times in the last 30 years. Only slightly more than half of the lawsuits resulted in damages being paid by the accountants. Auditors are most likely to be sued when a client commits fraud or declares bankruptcy.	Assign students to read about a recent lawsuit against an audit firm and describe the factors that led to the lawsuit.
Materiality	Read, Mitchell, and Akresh (1987)	Survey of 97 auditors from 24 accounting firms found low consensus on planning materiality judgments. Article includes suggestions for applying SAS No. 47 and evaluating the materiality of misstatements.	Give students the financial statements of a company and ask for their preliminary judgments of materiality. Discuss how differences in their materiality judgments could lead to different levels of testing.

(continued)

127

Appendix (Continued)

Topic	Research Studies	Major Findings	Classroom Exercises/Homework Assignments
Analytical Procedures	Bell, and Wright. (1997)	Auditors are prone to judgment errors when performing analytical procedures. Article includes suggestions for improving the effectiveness of analytical procedures.	Give students the situation described in Bell and Wright's *Exhibit 1* and see if the students consider management's unaudited balanced when forming expectations about the current year.
	Coglitore and Berryman. (1988)	Analytical procedures are effective for detecting revenue overstatements, fictitious sales and receivables, inventory overstatements, unrecorded liabilities, and other common misstatements. Article describes frauds that might have been detected if analytical procedures had been used properly.	Assign students to read a description of a fraud (e.g., ZZZZ Best, Miniscribe) and identify the analytical procedures that might have helped the auditors detect the fraud.
Evaluation of Going Concern	Geiger, Raghunandan, and Rama. (1996)	Less than 60% of bankrupt companies receive a modified audit report prior to bankruptcy. Large companies and clients of non-Big Six auditors are less likely to receive a modified audit report.	Ask students to suggest possible reasons why large clients are less likely to receive a going concern modification.
	Ponemon, and Raghunandan. (1994)	SAS 59 does not explicitly define substantial doubt. Auditors and financial statement users perceive substantial doubt as a high numerical probability, while judges and legislative staff members perceive substantial doubt as a lower numerical probability.	
SEC Disciplinary Actions Against Auditors	Brown, and Calderon. (1993)	Lack of independence and lack of skepticism are the most common criticisms cited in SEC disciplinary actions against auditors. Article describes common audit deficiencies and sanctions against auditors.	Assign students to read a recently issued Accounting and Auditing Enforcement Release and identify the auditing standards that were violated.

ACKNOWLEDGMENT

The author acknowledges gratefully the comments of Robert Sanborn, Bill N. Schwartz and two anonymous reviewers.

REFERENCES

American Institute of Certified Public Accountants. 1997. *Codification of Statements on Auditing Standards*. New York: AICPA.

Andersen, Arthur, Coopers & Lybrand, Deloitte & Touche, Ernst & Young, KPMG Peat Marwick, and Price Waterhouse. 1992. *The Liability Crisis in the United States: Impact on the Accounting Profession*. Statement of Position.

Bailey, C.D. 1986. Avoiding errors in judgment. *The Internal Auditor* 43 (June): 25-27.

Beaver, W.H. 1984. Incorporating research into the educational process. *Issues in Accounting Education* 2(Spring): 33-38.

Bell, T.B., and A.M. Wright. 1997. When judgment counts. *Journal of Accountancy* (November): 73-77.

Bell, T.B., and A.M. Wright, eds. 1995. *Auditing Practice, Research, and Education: A Productive Collaboration*. New York: American Institute of Certified Public Accountants.

Berton, L. 1996. Accountants expand scope of audit work. *Wall Street Journal* (June 17): B1, B3.

Brems, C. 1994. Taking the fear out of research: A gentle approach to teaching an appreciation of research. *Teaching of Psychology* 21 (December): 241-243.

Brown, P.R., and J.A. Calderon. 1993. An analysis of SEC disciplinary proceedings. *The CPA Journal* (July): 54-57.

Bryan, B.J., and L.M. Smith. 1997. Faculty perspectives of auditing topics. *Issues in Accounting Education* 12(Spring): 1-14.

Burilovich, L. 1993. Integrating empirical research into the study of introductory accounting. *Journal of Accounting Education* 10: 309-319.

Caster, P. 1992. The role of confirmations as audit evidence. *Journal of Accountancy* (February): 73-76.

Coglitore F., and R.G. Berryman. 1988. Analytical procedures: A defensive necessity. *Auditing: A Journal of Practice and Theory* 7(Spring): 150-163.

Dalton, D.R., J.W. Hill, and R.J. Ramsay. 1994. The big chill. *Journal of Accountancy* (November): 53-56.

Dopuch, N. 1989. Integrating research and teaching. *Issues in Accounting Education* 4(Spring): 1-10.

Geiger, M.A. 1994. The new auditor's report. *Journal of Accountancy* (November): 59-64.

Geiger, M.A., K. Raghunandan, and D.V. Rama. 1996. On serving the public's expectations: The ethical implications of SAS 59. Pp. 265-279 in *Research on Accounting Ethics*, Vol. 2, edited by L. Ponemon. Greenwich, CT: JAI Press.

Gersten, K. 1995. Debating the research paper. *Journal of Reading 39* (September): 71-72.

Girden, E.R. 1996. *Evaluating Research Articles from Start to Finish*. Thousand Oaks, CA: Sage.

Hermanson, R.H., P.H. Duncan, and J.V. Carcello. 1991. Does the new audit report improve communication with investors? *The Ohio CPA Journal* (May-June): 32-37.

Houghton, C.W., and J.A. Fogarty. 1991. Inherent risk. *Auditing: A Journal of Practice and Theory* 10(Spring):1-21.

Joseph, G.W., R.M. Keith, and D.R. Ellis. 1996. Understand your privileges and responsibilities under the copyright law. *Issues in Accounting Education* 11(Spring): 77-82.

Leisenring, J.J., and L.T. Johnson. 1994. Accounting research: On the relevance of research to practice. *Accounting Horizons* 8(December): 74-79.

Loebbecke, J.K., M.M. Eining, and J.J. Willingham. 1989. Auditors' experience with material irregularities: Frequency, nature, and detectability. *Auditing: A Journal of Practice and Theory* 9(Spring): 1-28.

Lowe, D.J., and K. Pany. 1993. Expectations of the audit function. *The CPA Journal* (August): 58-59.

Markham, W.T. 1991. Research methods in the introductory course: To be or not to be? *Teaching Sociology* 19(October): 464-471.

Oliverio, M.E. 1984. The audit as a scientific investigation. *The CPA Journal* (October): 52-60

Palmrose, Z.V. 1997. Who got sued? *Journal of Accountancy* (March): 67-69.

Pany, K., and S. Wheeler. 1992. A test of analytical procedure effectiveness. *The CPA Journal* (June): 64-67.

Pany, K., and O.R. Whittington. 1997. *Auditing*, 2nd ed. Chicago: Richard D. Irwin.

Ponemon, L.A., and K. Raghunandan. 1994. What is "substantial doubt"? *Accounting Horizons* 8(June): 44-54.

Pratt, J., and J.D. Stice. 1994. The effects of client characteristics on auditor litigation risk judgments, required audit evidence, and recommended audit fees. *The Accounting Review* 69(October): 639-656.

Public Accounting Report. 1991. Lowballing: The quiet war in audit fees. (September 30): 1, 4.

Read, W.J., J.E. Mitchell, and A.D. Akresh. 1987. Planning materiality and SAS No. 47. *Journal of Accountancy* (December): 72-79.

Sterling, R.R. 1973. Accounting research, education and practice. *Journal of Accountancy* (September): 44-51.

Wright, S. 1997. Bringing research into the classroom. *The Auditor's Report* 21(Fall): 16-17.

ANALYZING AN INTERNATIONAL ANNUAL REPORT AS A COURSE PROJECT

Robert Bloom and David Schirm

ABSTRACT

This paper offers advice to college faculty on assigning an international annual report project in an advanced undergraduate or graduate course, especially the international accounting course. The rationale for this assignment is that it serves to apply many of the concepts and practices covered throughout the course and in prerequisite courses to a particular country in a comprehensive manner.

This paper also furnishes a comparative analysis of various aspects of computer databases that include annual report data. Thus, this paper provides the advantages and disadvantages of selected databases to use in conducting financial report projects, whether the company is based in the United States or abroad.

Advances in Accounting Education, Volume 2, pages 131-163.
ISBN: 0-7623-0515-0

INTRODUCTION

In the international realm, the Accounting Education Change Commission (AECC) called for "international and multicultural knowledge" as part of the general education for accountants. The AECC asserted that accountants' knowledge should include "an awareness of the different cultures and socio-political forces in today's world..." (1990). The AECC also observed that accounting knowledge should include... "[i]n-depth knowledge in one or more specialized areas, such as...international accounting." Earlier, the Bedford Report recommended that "[a]ll college graduates...have a strong, broad general education," including "international and multicultural experiences" (1986, 180). The AECC and Bedford Report assertions in the international realm provide the motivation for this paper.

The primary aim of this paper is to provide advice to college faculty on assigning an international annual report project in an advanced undergraduate or graduate course. This paper stems from the frustration one of the co-authors experienced in teaching a graduate course in international accounting and assigning a financial report project to each student in that course. The rationale for this assignment is that it serves to apply many of the concepts and practices covered in this course and prerequisite courses to a particular country. Therefore, the instructor can give this assignment to the class in lieu of a final examination. Another aim of this paper is to offer a comparative analysis of various aspects of computer databases to use in conducting financial report projects, whether the company is based in the United States or abroad.

PREVIOUS LITERATURE

In the recent accounting education literature, there are no articles on annual report projects of the kind that we describe in this paper. Nevertheless, there are a few relevant articles.

Tondkar, Adhikari, and Coffman (1994) discuss two approaches to internationalizing upper-level financial accounting courses: (1) integrating selected topics into particular traditional courses, and (2) developing a separate international course. The authors furnish annual reports for eight international companies they use in implementing both teaching approaches. The authors also offer a comparative analysis of accounting topics and their treatment in the United States, Canada, Australia, United Kingdom, France, Germany, the Netherlands, and Japan including the relevant pages, if any, in the specimen annual reports from each country pertaining to the topic (e.g., goodwill amortized, financial leases capitalized).

O'Connor, Rapaccioli, and Williams (1996) provide three steps for internationalization of the advanced accounting course, which Stout and Schweikart (1989) found in a faculty survey to be the course to internationalize if internationalizing is limited to only one course. The three steps are: (1) select a few developed countries in order to contrast U.S. GAAP with their counterparts—e.g., consolidations, business combinations, foreign currency transactions and translations. (2) introduce early in the course the factors underlying differences in accounting from one country to another. (3) use examples from non-U.S. financial statements to emphasize differences between U.S. and other GAAP.

Wallace (1996) provides ratios for Germany, Italy, Japan, and the United Kingdom, the countries on which O'Connor, Rappaccioli and Williams (1996) focus, using a Disclosure International database. Various accounting practices affect ratios as do reconciliations with U.S. GAAP. Instructors can use the ratios in class in terms of the following question Wallace raises (337): "How do observed ratio variations relate to described differences in accounting systems?" [Italics in the original]

Our study offers faculty in advanced courses ideas on how to internationalize such courses by having their students analyze international financial reports. In the course of this analysis, students must become acquainted with the culture of the country in which the company is located as well as its GAAP. Students use the foregoing information to evaluate the performance of the company with a view to deciding whether to invest in this company.

THE COURSE

Our university has offered an international accounting course once a year since the Spring 1997 semester. Thus far, only M.B.A. students have enrolled. The M.B.A. program has been a generalized course of study without majors. Starting in the summer of 2000, this program will offer an accounting concentration as part of a five-year, 150 hour course of study. We then expect the students concentrating in accounting to take the international course as either graduate or undergraduate matriculants in the program.

The course emphasizes the linkage between culture and accounting in different countries. In class, we discuss issues and controversies in international accounting. Students prepare analyses of case studies each week.

In teaching the international accounting course, we attempt to cover the following topics in this order: comparative accounting principles and cultural underpinnings, accounting under inflationary conditions, foreign currency translation, and foreign currency transactions. (Appendix A provides the current syllabus for this course.) By covering international accounting principles and their cultural foundations early in the course, students have the required background to do the term papers.

The class size has varied from six to ten students a semester, most of whom were not undergraduate accounting majors. Our experience in the course suggests that the few students with accounting backgrounds who have taken the course do not perform as well as the other students because they are less able to adapt to the case method and analyze unstructured accounting situations. Also, the "accounting" students are less skillful than the others in preparing the term papers.

We require students to prepare two term papers: The first is worth 20% of the course grade. It is an analysis of a country, including the development of a conceptual accounting framework and a discussion of its main generally accepted accounting principles. We require a comparison of those principles with U.S. GAAP, along with explanations of similarities and differences in light of cultural, historical, economic, and legal aspects of the two countries. The second paper is worth 35% of the course grade. It is an analysis of the performance of a non-U.S.-based company, primarily, but not exclusively, from an examination of its annual report data. Also relevant to this analysis is an examination of news media articles and industry periodicals about this company. While the students have several options for the second term paper, almost all of them select the option just described, which is the subject of this paper. Furthermore, in pursuing this option, students also have to do the same work they previously did in the first term paper, but this time applied to a company in another industry and country.

This project requires the following components: (1) background on the company, (2) consideration of the country and culture in which the company is headquartered, (3) development of a conceptual framework for the country, (4) analysis of generally accepted accounting principles in this country in light of the culture and conceptual framework, followed by (5) analysis and evaluation of the performance of the company using ratio analysis.

Students compare the ratios of their company with those of similar companies in the same industry from the same country and other countries. The key problem with the ratio analysis is that industry ratio norms, while widely published for U.S. companies (see, as an example, *Troy's Almanac* 1999), are not available for other countries. Only the Center for International Financial Analysis and Research (CIFAR 1993) has provided industry ratio norms by country, but this service has been discontinued.

The question is how students can do a meaningful ratio analysis today. One approach that some of them have used is to resort to CIFAR's 1993 database, which is becoming dated. Students also have developed their own industry ratio norms for all intents and purposes by computing the same ratios for a group of similar companies in much the same industry within the same or similar countries from consecutive annual report data.

Securing the original annual reports of a multinational company based abroad can be a difficult task. Libraries and universities often subscribe to comprehensive computer databases such as: (1) Standard & Poor's Compustat, which includes PC Plus for U.S. companies and foreign companies listed on U.S. exchanges reporting in U.S. GAAP, and Global Vantage, which includes foreign and U.S. companies reporting in their national accounting standards; (2) Moody's International Company Data; and (3) Disclosure, produced by Primark, which offers

Worldscope, among others. These databases, in varying degrees, facilitate ratio analysis.

Steps In The Analysis

In preparing their studies, the students use a common format, consisting of five steps. This format facilitates their presentation and our evaluation of these studies. The analysis of the company includes its background, fundamental accounting framework, conceptual framework, and generally accepted accounting principles as well as an examination of its financial statements. The students follow these steps:

- In the first step, they describe the overall background of the country—its economy, legal traditions, social customs.
- In the second step, they analyze the nature of the accounting framework in the country in which the company is headquartered.
- In the third step, they develop a conceptual framework of financial reporting for the country.
- In the fourth step, they apply the background of the country and its conceptual framework to analyze and evaluate the key generally accepted accounting principles in the country.
- In the fifth step, they prepare a financial statement analysis of the company.

The *first step* considers the background of the country—its history, the nature of its economy, the nature of its legal system, and its culture (i.e., its customs and values)—as well as the background of the company. Students must analyze the culture in terms of the Hofstede and Gray models. Hofstede (1980, 92-341) derives cultural values from a survey of employee attitudes in 50 countries:

1. *Individualism versus collectivism*—a preference for individual responsibility, emphasizing the individual rather than the group.
2. *Power distance*—the extent to which people in a country are willing to accept unequal distributions of power in institutions and organizations.
3. *Uncertainty avoidance*—the extent to which people cannot tolerate uncertainty and ambiguity.
4. *Masculinity and femininity*—the extent to which a mindset exists in the country emphasizing the importance of self-assertiveness, the need to achieve, if not be a hero, versus the opposite mindset involving a concern for quality of life, caring relationships, and sympathy for the poor and ill.

Gray (1988, 8-14) derives the following four accounting attributes from Hofstede:

1. *Professionalism versus statutory control*—a preference for exercise of individual professional judgment and self-regulation in contrast to adherence to legal requirements and statutes.

2. *Uniformity versus flexibility*—a preference for standardized accounting practices among companies and organizations as opposed to flexibility based on different accounting circumstances or conditions.

3. *Conservatism versus Optimism*—a tendency in accounting when alternatives exist to reflect the least favorable immediate effect on net income and wealth. Conservatism constitutes a biased, yet prudent approach to measuring accounting wealth.

4. *Secrecy versus transparency*—a penchant for confidentiality in financial reporting to avoid disclosing information to competitors and the public at large.

The **second step** deals with the nature of the accounting framework in the particular country. Worldwide, there are at least five such frameworks (Mueller et al. 1997, 12-13; Holt and Hein 1999, 6-14): Anglo-Saxon, Continental, Latin America, Mixed Market, and Islam.

The Anglo-Saxon framework applies to primarily English-speaking countries (such as the United States, Canada, the United Kingdom, Australia, and New Zealand), having a common law system, but also to Hong Kong, Indonesia, Israel, India, Malaysia, Netherlands, and the Philippines, among other countries that were originally British colonies or U.S. protectorates. In this framework, investors are the primary and creditors the secondary users of financial reports. The stock market provides the main source of capital. Accordingly, nongovernmental organizations develop the accounting principles with the government overseeing their work. Tax regulations and accounting principles are different. To accommodate a vast array of investors and creditors, financial reports provide extensive disclosure.

In the Continental framework (including countries such as France, Germany, Ivory Coast, Japan, Morocco, Portugal, Spain, and Sweden, among other countries that were originally colonies or protectorates of the foregoing), the principal users of financial reports are creditors (e.g., bankers), the government, and investors, in that order. Primarily banks provide the capital in these countries. Such countries adopting this framework use a code law system. The law is the basis of accounting principles, formulated by a governmental agency. Tax regulations and accounting principles are similar, if not the same. Financial reports do not provide full disclosure because the users of these reports have little difficulty securing the information they need from the companies. Financial reporting emphasizes conservatism because this is what bankers desire.

The Latin American framework parallels the Continental model since Spain and Portugal, which belong to the Continental framework, established colonies in Latin America. There are two notable differences between the two frameworks. In the Latin American model, in addition to creditors and the government, the main users also include wealthy landowners. Also, in view of significant inflation in some of these countries, financial reporting may require historical cost adjusted for a constant dollar although current replacement cost accounting can often supplement (if not supplant) the general price level, historical cost data.

Traditionally, governmental centralized planning characterizes the Mixed Market framework, which applies to the former Soviet Union and its Commonwealth of Independent States as well as China. The government is the user in the socialist version of this model. There is no external financial reporting. Therefore, the extent and quality of disclosure in financial accounting is a non-issue. The government prescribes accounting consistent with the tax law using a uniform chart of accounts. The Ministry of Finance establishes accounting principles. In view of their code law systems, accounting principles follow legal developments closely.

The capitalist version of the Mixed Market framework emphasizes privatization of industrial enterprises by the government in order to raise capital through foreign investment and loans. The main users of accounting information are investors and creditors. With privatization, there is a need to develop generally accepted principles that can be applied and interpreted uniformly. The scope and quality of disclosure in annual reports is imperative. The need for relevant and reliable financial reports is evident.

In Islamic countries, religion affects all aspects of life, including business and accounting. Interest for using money is illegal in these countries, which have code laws. Instead of interest, the lender may receive shares of stock in the company. Since interest is forbidden, there are no applications of discounted cash flows in accounting. Each business firm must make charitable contributions. The government and banks furnish the principal sources of business financing and use accounting reports in credit decisions. Financial reports offer limited disclosures since the main users have no difficulty securing the data they desire directly from the company itself. Few generally accepted accounting principles exist. There is no accounting regulatory body, and the accounting profession is weak in Islamic countries. Accounting is conservative with accelerated depreciation methods predominantly used and with income smoothing a popular practice (Kantor et al. 1995, 31-50).

The ***third step*** is to develop a conceptual framework of financial reporting for the country, including objectives and qualitative characteristics. As a starting point, students use the FASB's framework, including the key objectives and qualitative attributes. The first objective set forth in this conceptual framework is to convey information in financial reports useful to particular users (e.g., investors) and for specific purposes of the company (e.g., decision-making). The second objective is to provide information in the reports to help those users achieve the first objective. The third objective is to provide information about the company's financial resources (its assets), the claims to those resources (its liabilities), and the changes in the company's financial resources and claims to those resources, again to facilitate the first objective.

The *fourth step* applies the background and conceptual framework sections to analyze and evaluate the main generally accepted accounting principles in the country. Do those principles follow from the background and conceptual framework? If that is not the case, students provide an explanation.

The *fifth step* entails a financial statement analysis. In the financial statement analysis, students compute ratios for at least three years, preferably five. If available, they can use foreign industry norms. If norms are not available, then the students should perform the ratio analysis for several other companies in the same industry, preferably in the same or similar countries. We require this analysis in terms of the GAAP used by the company. More specifics on the financial statement analysis using the databases appear in the next section of the paper.

The financial statement analysis includes an assessment of firm risk and market risk. As discussed below, both Moody's and Compustat have default ratings. Compustat has betas while Moody's and Disclosure do not. Additionally, Compustat provides monthly stock and market index returns data in contrast to Moody's and Disclosure. The student should compute betas, if they are not available, assuming that company and market returns are reported.

Finally, as part of the financial statement analysis, students analyze the capital stock of the company to decide whether to purchase shares of the company's stock. The students must provide a complete explanation of their choice based on their work from the first four steps.

Financial Statement Analysis

Techniques for analyzing the historical financial performance of a business organization include calculation of financial performance ratios from summary financial statements. Students evaluate these ratios by comparing their values over time and with those of similar business organizations. For U.S. corporations, standard sources of industry ratio data in book format include *Industry Norms and Key Business Ratios* (Dun & Bradstreet 1999), *Almanac of Business and Industrial Ratios* (Troy 1999), and *Annual Statement Studies* (Robert Morris Associates 1998). *The Global Company Handbook* provides similar information for non-U.S. corporations. Additionally, in recent years, computer databases of historical financial statement data have allowed users to collect financial statement data for selected firms. The data permit the users to calculate average ratio values for selected firms using either report programs included in the software or to download the required information to spreadsheet software for ratio calculations. Standard and Poor's, Moody's, and Disclosure Incorporated, among others, have developed such databases. Computer financial statement databases have evolved from computer tapes to CD-ROMs to Internet sites.

Table 1. International Financial Electronic Databases

	Compustat PC Plus (PC) & Global Vantage (GV)	Moody's International Company Data	Worldscope Disclosure
Company Coverage	PC: 19,000 companies U.S. and Canadian , GV: 12,000 companies U.S. and abroad	63 U.S. companies and 11,867 non-U.S. companies	15,900 U.S. and abroad
Time Period	PC: Latest 20 years , GV: 1982—present	Latest 5 years	Latest 10 years
Financial Statement Format	Multiple standardized reports (150)—some in spreadsheet format	Local country and cross-country chart of accounts—can be saved in spreadsheet format	Single standard format— can be saved in spreadsheet format
Statement Footnotes	Information from foot-notes, not the original footnotes	Extensive debt structure information	Most recent annual report in Adobe file
Annual Report	Not available	Not available	Most recent
Sec Filings	Not available	Not available	Listing and links to individual filings
Ratio Analysis	Three company annual ratio reports—industry average ratios with profitability report	Company ratio report for most recent year includes peer group average ratios	Company ratio report for last three years—no peer group average ratios
Risk Analysis	S&P default grades; estimated equity betas; historical monthly equity returns—company and index	Extensive information on company default risk and debt structure; no beta	No default or beta
Earnings Forecasts	Not for university subscribers	Not available	Available for an additional fee
Update Frequency	Subscription dependent— annual for universities	Quarterly	As new information becomes available
Software Manuals	Extensive documentation	Limited	None
Vender Assistance	Extensive 800 telephone assistance	Limited 800 telephone assistance	E-mail assistance, online Help
Ease of Use	Requires significant user product specific knowledge	Easy	Very easy
CD/Internet	CD—single site or network	CD—single site or network	Internet
Academic Research Use	Extensive—widely used in academic business literature	Very limited use	Very limited use
Miscellaneous	Comprehensive report creation and search features; over 1000 searchable data fields	Strong on debt disclosure	Comprehensive current company information with Internet links
Academic Price (Annual)	PC: $9,000, GV:$10,000—single site, additional network charge	$4,000—single site, additional network charge	$11,175—single simultaneous user, additional user charge

For non-U.S. corporations, Compustat PC Plus and Compustat Global Vantage, Moody's International Company Data, and Disclosure's Global Access international financial database furnish historical financial statement data in computer database format. These CD or Internet database products differ in various dimensions. They include: number of firms covered, the number of individual summary accounts for each firm, the report format of presentation of financial data for different types of business organizations in different countries, and the number of years of historical information included in the database. Each computer database comes with preprogrammed reports and the capability for customized report creation, using the fields available from the original database. See Table 1 for a comparison of these database products.

Compustat provide two database products with financial statement data for international companies: PC Plus and Global Vantage. PC Plus includes foreign firms that trade on U.S. exchanges and provides financial statement information based on U.S. GAAP. PC Plus also includes Canadian firms that trade on Canadian exchanges. The 1998 version of PC Plus contains information on more than 10,000 active companies as well as another 9000 companies no longer in business or operated as public companies. PC Plus provides annual historical data for as many as 20 years for each company with over 800 data items possible for each year.

Global Vantage presents financial statement information based on the national GAAP of each firm in the database. The 1998 version of Global Vantage includes financial statement data for approximately 12,000 companies from 70 countries beginning with financial data for as early as 1982. The database includes separate report formats for industrial firms and various types of financial business organizations.

PC Plus and Global Vantage provide 150 different report formats. The student can download data from these reports in text or spreadsheet format for subsequent analysis. Global Vantage reports allow different data presentations for different countries and different types of business organizations. Both databases contain three ratio reports: a five-year report, a comparative report, and a summary report. The five-year report furnishes 30 different ratios, measuring liquidity, activity, profitability, and leverage for a user-specified, five-year period. Similarly, the comparative annual ratio report provides the same ratios for up to six separate companies for a specified one-year period. (Appendix B provides examples of these reports.) The summary report furnishes 39 different ratios, measuring profitability, market valuation, leverage, and turnover for a specified ten-year period as well as period averages and growth rates for these ratios. The student can download these reports as text or spreadsheet files, thus allowing the calculation of average ratios for a particular set of firms. In addition, Global Vantage provides eight preprogrammed charts, which compares a specific ratio for a designated firm with a user-defined peer group average for the same ratio over a five to ten year period. Students can print the reports, which support these charts. The software has a flexible report creation feature, whereby students can specify group ratios for a specified set of firms.

The 1998 version of Moody's International Company Data includes the latest five years of financial data for 63 U.S. and 11,867 non-U.S. companies. Moody's International Company Data offers a choice of two standard reports for financial data, utilizing either a home country or a foreign chart of accounts. Students can apply a peer group ratio analysis for a specified sample of firms from the overall database. This peer group feature calculates eleven ratios as well as eight aggregate balance sheet and income statement accounts. (See Appendix C for examples of the Moody's report.) The ratios measure profitability, activity, liquidity, and leverage. The peer group analysis permits a comparison of each company in the specified group to the average ratio of other companies in the group, the same country, and the same region. For the most recent fiscal year, all ratios are available, while for all years in the database summary ratio component data are available. Thus, students can define a report that would include the summary accounts necessary for the selected group of firms for all years in the database. Students can extract the data in a format that allows for further calculation of the ratio values using spreadsheet software. No market risk data and analysts' earnings forecasts are available in the database. There is more extensive information on each company's debt structure consistent with Moody's debt evaluation services.

Disclosure Global Access Internet (Disclosure) provides access to historical financial statement data for both U.S. and international firms. The database includes ten years of historical data for over 15,000 international firms from more than 50 developed and emerging markets. Disclosure furnishes financial statement data in a standard report that includes a section of some 29 fundamental ratios covering profitability, asset utilization, leverage, and liquidity for the past three years. The standard report does not provide peer group ratio data, but students can construct such ratios from financial statement data of individual companies downloaded to a spreadsheet. Also, Disclosure is developing an information application for the database called Piranha to include peer analysis features. This database currently includes analysts' earnings forecasts. (See Appendix D for an example of a Disclosure report.)

The amount of specific user information necessary to utilize each database varies, with the Disclosure product requiring only general familiarity with accessing information on the Internet. The Moody's and Compustat Global Vantage products require users to have more specific database information. Moody's provides a much more limited number of data fields, screening capabilities, and report writing capabilities. Thus, this database requires less specific information than the more comprehensive and flexible Compustat Global Vantage Product, developed initially for use by professional financial analysts. COMPUSTAT improved its user friendliness by replacing the DOS based procedures with a friendlier Windows environment. COMPUSTAT also offers the most support for users of its products, including extensive written documentation, extensive help information in the database, and 800 telephone service for all authorized subscribers. Both the Moody's and Disclosure products offer much more limited written documentation, help information in the database, and telephone support.

All three database vendors offer academic institutions discounted prices for their products compared to commercial accounts. Table 1 furnishes a representative annual price for each product as a single purchase. When universities purchase other products from the same vendor, they can obtain further price discounts. Both the Compustat and Moody's products require additional university expenses for computer hardware and support. Utilization of Disclosure's Internet product simply requires a valid user name and password and a computer with Internet access, thus minimizing the need for additional university resources beyond the subscription price of the software. Moody's soon will make the International Company Data product available for Internet subscription.

CASE EXAMPLE

Students select a foreign company to analyze following the five steps outlined above. The analysis below is a case example that we use as part of our strategy of teaching international accounting. As a result of this analysis, students assess the performance of the company and decide whether to invest in it.

We selected Daimler-Benz (D-B) as our example. The first German company to be listed on the New York Stock Exchange in 1993, D-B has had to comply with SEC regulations since that time. In 1996, D-B decided to switch entirely to U.S. GAAP.

In the *first step* of the analysis, we consider Germany—its history, economy, legal system, and culture—as well as background on D-B. A member of the European Union, Germany has a code law system. The country is, in many respects, a welfare state with high labor costs. Germany belongs to the Continental accounting framework. In Germany, tax regulations and accounting principles are essentially the same. Legislation and court decisions represent the main source of accounting principles. The German Commercial Code sets forth such principles or regulations. Thus, one set of financial statements exists for each company, and few, if any, timing differences between accounting and taxation occur. Therefore, companies typically do not report deferred taxes, apart from preparing consolidated financial statements through elimination entries. To achieve legal compliance, auditing is the primary concern of professional accounting organizations, not accounting as such since that is dictated by law (Choi et al. 1999, 58).

Since D-B listed its shares on the New York Stock Exchange in 1993, it has improved the efficiency of its business operations. Profitability serves to enhance the value of its capital stock and thus benefits the firm in raising funds in the United States and elsewhere. D-B was excessively diversified, but is now firmly focused on motor vehicle production and sales.

The company has long had a bureaucratic organizational structure. If it can reduce the size of its work force as Chrysler has done, profitability can improve. Furthermore, the firm can reduce costs significantly by moving production and assembly plants out of Germany (e.g., the ML sport-utility vehicle assembly plant in Tuscaloosa, Alabama). Most of D-B's production facilities are still located in Germany, where the wages are high and the productivity is low (Avila 1998).

Chrysler runs a more efficient operation. It takes only two years for this company to introduce a new car, starting with the design stage (Avila 1998). Operating on an informal basis, Chrysler has a lean staff. In recent years, Chrysler has earned more on a $20,000 car than D-B on a $43,000 car (Avila 1998).

The *second step* in the analysis deals with the nature of the accounting framework in Germany. This country follows the Continental framework. Banks and creditors are the principal users of financial statements in Germany, followed by the government, investors, and employees. Banks are the major source of corporate financing. Since banks own major corporations and CPA firms as well, and have no difficulty securing the information they need directly from the companies, financial reports provide minimal disclosure. The law requires that companies set up legal reserves, putting aside a portion of retained earnings to preserve capital and protect the creditors. Companies also may use hidden or secret reserves (i.e., undervaluation of assets, overstatement of expenses and liabilities) to smooth income and protect creditors (Choi et al. 1999, 92). Historical cost is the valuation method.

In the *third step*, we formulate a conceptual framework of financial reporting. Germany has no such explicit conceptual framework of financial reporting. If we were to develop such a framework of objectives, it could be to provide information:

1. of a stewardship nature, in terms of the resources and debts of the company and changes in the resources and debts during the year;
2. of a financial nature, to assist banks, other creditors, the government, and to a lesser extent investors and employees in making economic decisions about the firm;
3. of an historical nature, to assist the users in forecasting the future cash flows of the firm in terms of amounts, timings, and the uncertainties associated with the amounts and timings.

The qualitative characteristics of German financial reporting include: conservatism, reliability, objectivity, prudence, comparability, and consistency.

In the *fourth step*, we apply the background and conceptual framework sections to Germany's GAAP. In specific terms, to appeal to bankers and the government, Germany emphasizes historical and conservatism in financial reporting. Companies cannot reflect holding gains on fixed assets. In Germany, a code law country, income tax regulations determine depreciation on these assets. The application of the lower-of-cost-or-market rule and LIFO (if the physical flow of goods is

Table 2. Comparing German and U.S. GAAP

	Germany	U.S.
Legal System	Code Law	Common Law
Purpose of financial statements	Stewardship	Users' decision making
Tax regulations versus accounting principles	Essentially the same	Different
Disclosure in financial reports	Limited	Detailed
Principal users of financial reports	Bankers, other creditors, and the government	Investors and Creditors
Historical cost	Greatly emphasized	Emphasized
Conservatism	Greatly emphasized	Not so much emphasis
Contingent Losses	If possible and reasonably estimable, journalized[*]	If probable and reasonably estimable, then journalized
Discontinued operations	No reporting or disclosure required	Required reporting in income statement
Extraordinary gains and losses	Footnote disclosure	Required reporting in income statement
Financial leases	Expensed	Capitalized
Foreign Currency Translation Method	No single required method	Current Rate Method Predominant
LIFO	Acceptable if Physical Flow of Goods LIFO, but Seldom Used	Widely Used
Long-term Construction Contracts	Completed Contract Method	Percentage of completion or Completed Contract Methods
Pensions and post retirement benefits	Not journalized or disclosed[**]	Journalized and disclosed
Research and Development Costs	Generally expensed as incurred	Generally expensed as incurred
Segmented disclosures	Segments reported only for revenues by industry and geographic markets	Segmented according to way firm is actually managed, e.g. product line, geography, customers; reporting revenues, income, identifiable assets.
Statement of Cash Flows	Not required	Required

Sources: [*]Iqbal et al. (1997, 75).
              [**]Iqbal et al. (1997, 226).

approximated by this method) to inventory valuation illustrate conservatism. Also, most companies write off goodwill immediately, though some firms amortize it over four to 20 years (Choi et al. 1999, 58). Also, in conformity with conservatism, companies expense research and development costs as incurred. The common practice is to expense financial leases as companies rent rather than capitalize the lease on the balance sheet. Companies typically apply the temporal method of foreign currency translation, showing exchange rate losses in the income statement. Financial statements do not reflect unrecognized gains on translations or any other event. Firms report provisions for future expenses and losses as soon as possible in order to smooth income. With respect to long-term construction contracts, companies can only use the completed-contract approach.

Based on our analysis of its background and conceptual framework, Germany's generally accepted accounting principles appear to be compatible. These principles reflect the conservative practices desired by bankers and the government. (In the course of analyzing German GAAP, some students prefer to compare German and U.S. GAAP. For such a comparison, see Table 2.)

The *fifth step* entails a financial statement analysis. For this ratio analysis, we decided to use Compustat PC Plus and Global Vantage databases to provide a variety of reports in standardized spreadsheet format. Of the four databases we considered, PC Plus and Global Vantage are the most widely used by financial analysts and academics alike. In fact, at least 59% of America university business schools that are accredited by the American Assembly of Collegiate Schools of Business subscribe to Compustat products like PC Plus and Global Vantage (Lenox 1998, 41).

Each database provides the same ratio reports, although PC Plus and Global Vantage report the basic financial statement data in U.S. GAAP and national GAAP, respectively. (See pp. 138-142 of this paper for a description of these databases.) By contrast, Disclosure furnishes ratios for individual companies, which are downloaded and integrated for the separate companies into a spreadsheet format for comparative analysis among the peer group. Moody's CD International database currently does not provide a ratio analysis for a peer group, country, or region. Moody's only offers a ratio analysis for an individual company for the most recent annual report, making its ratios more limited than the Disclosure product.

Ratios provide clues, not answers, to questions about the operating and financial performance of the enterprise. Only through investigation can the clues yield answers. For ratios to be useful, one should compare them with the corresponding figures from previous financial statements as well as those for the principal industry in which the firm is situated.

As our initial peer group for D-B, we used five other German automobile companies: Adam Opel,[1] Audi, BMW, Ford-Werke,[2] and Volkswagen. Our comparative evaluation of their ratios follows. This peer group evaluation pertains to 1993, 1994, and to 1995 and uses data from Compustat Global Vantage. For the 1996 and 1997 ratio evaluations, since D-B began to apply U.S. GAAP exclusively starting in 1996, the peer group we selected consisted of Chrysler, Ford, and General Motors in the U.S. (the Big-Three). The definitions of the ratios for the German automobile and Big-Three companies are the same. The D-B and Big-Three ratios for 1996 and 1997 are taken from the Compustat PC Plus. Both Global Vantage and PC Plus reflect the differences in German and U.S. GAAP.

For instance, Germany includes cash surrender value of life insurance among other current assets in contrast to U.S. GAAP. Also in Germany, companies translate receivables in foreign currencies at the lower of the closing rate and the historical rates; payables at the higher of those rates. Companies tend to reflect translation losses, but not translation gains, in income (Alexander and Archer 1992, 247).

From 1993 to 1995, using the German peer group, the current ratio D-B shows 10% higher and more stable figures than its counterparts. For the quick ratio, D-B reflects a 5% higher set of figures than the other German companies apart from Audi. Thus, D-B appears to be more liquid than most of its German peer companies.

For 1996, D-B's current ratio exceeds the Big-Three average by 20% and especially Chrysler's figure. In 1997, D-B's current ratio continues that pattern—30% higher. The quick ratio for D-B in 1996-1997 mirrors the Big-Three average, but is considerably above Chrysler's figure—33% higher in 1996 and 50% higher in 1997. Thus, Daimler's liquidity is evident from 1993 to 1997.

With respect to inventory turnover, D-B shows a stable, but low set of figures relative to its German peer group. The inventory turnover ratio for D-B in 1996 is about 49% of Chrysler's figure, and 38% of the Big-Three average. In 1997, D-B's inventory turnover improved, but represents approximately 67% of Chrysler's figure and 47% of the Big-Three average. Thus, the company's inventory turnover ratio needs improvement.

D-B's receivables turnover is generally lower than its German peers, declining from 1993 to 1995: 41% of the peer group average in 1993, 26% in 1994, and 18% in 1995. This figure is significantly above the Big-Three average for 1996 and 1997, 30% and 31% higher, respectively, though somewhat below Chrysler's figures for these periods, 80% of Chrysler's 1996 figure, 82% in 1997. The asset turnover is among the lowest of the German peer group: 64% in 1993, 59% in 1994, and 55% in 1995 of the peer average. This figure is 20% and 24% higher, respectively, in 1996 and 1997 than those of the Big-Three, but close to Chrysler's amounts for those years. D-B's collection period is the highest by far and from 1993 to 1995 for the German companies: 78%, 126%, and 158% higher than the peer average. The average collection period, while high for the German group, is nearly one-half of the Big-Three average in 1996 and 1997, though somewhat higher than Chrysler's figures—26% and 22% higher than Chrysler's ratios. The days-to-sell inventory is high relative to the other German companies. This figure is two and one-half times as high as the Big-Three's average in 1996, twice as high in 1997. Compared to Chrysler, the days-to-sell inventory is nearly twice as high in 1996, one and one-half times as high in 1997. D-B needs to improve its days-to-sell inventory figure.

This company has the longest German operating cycle apart from Ford-Werke. The operating cycle, which improved from 1996 to 1997, is considerably below the Big-Three average figures—79% of 1996 and 69% of 1997, but higher than

Chrysler's figures for these years—48% higher in 1996 and 29 higher percent in 1997. Thus, the activity ratios reflect mixed results.

As for Revenues/Net fixed assets, this ratio is the lowest, except for Volkswagen, which is slightly below D-B, casting a cloud over its efficiency. This ratio is above the Big-Three average for 1996-1997, 7% and 10% higher in those two years, and 12% and 34% above Chrysler's figures in the same period. With respect to Revenues/Stockholders' Equity, this ratio by far is the lowest among the German firms: 44% in 1993, 49% in 1994, and 42% in 1995 of the average. This figure also is significantly below the Big-Three average—69% in 1996 and 54% in 1997 and Chrysler's corresponding figures—79% and 69%.

The negative profitability measures of four of its major competitors complicate a comparison of D-B's positive profitability with the corresponding figures of its German peers in 1993. In 1994, D-B's pretax profit margin is essentially equal to its peers. In 1995, D-B took a "big bath" with a significant net loss, while its German peers all reported positive profitability measures for the same year. The pretax margin is considerably below the corresponding ratio for the Big-Three in 1996-1997, 29 and 53%, respectively. Compared to Chrysler, this figure is 18% of the 1996 and 44% of the 1997 figures. The net profit margin also is low. This figure is far below the Big-Three average and Chrysler's figure for 1996—62% and 42%, respectively, with a dramatic change in 1997—147% and 136%, respectively, considerably above the Big-Three average and Chrysler's figure. For return on assets, D-B reflects higher income relative to assets than its German peers, but Adam Opel is the best of this group. The return on assets is 70% of the Big-Three average and 38% of Chrysler's figure in 1996. The pattern reverses in 1997: D-B is 76% higher than the Big-Three average, but only 26% higher than Chrysler. Return on common equity is clearly not the best, but more or less average for the German companies. D-B's 1996 figure is 45% of the Big-Three average and 33% of Chrysler's figure. A dramatic improvement occurred in 1997 with 81% of the Big-Three average and 93% of Chrysler. D-B's return on common equity needs to be improved.

D-B's interest coverage before tax clearly is not best among the German companies. Audi's coverage ratio is the highest, with the exception of 1993. Compared to the Big-Three in 1996, interest coverage for D-B is 93%, though only 50% of Chrysler's figure. In 1997, D-B's figure is 68% higher than the Big-Three average, though only 6% higher than Chrysler's figure. Interest coverage before tax needs improvement.

As for long-term debt/common equity, D-B shows significant financial leverage, but only one-third of Volkswagen, relative to the German automobile companies, one-fourth of the Big-Three average for 1996-1997, yet 73% and 62% of Chrysler's figures for 1996-1997. Total debt/total assets appears to be a stable and consistent ratio for D-B, though low relative to BMW and Volkswagen, and considerably lower than the figures for the U.S. peer group, but close to Chrysler's figures in 1996-1997. Total assets/common equity is lower for D-B relative to its

German peers with the exception of Audi in 1993 and 1994. For this ratio, compared to the Big-Three, the 1996 D-B figures are more than half the size, but similar to Chrysler's figures; for 1997, 40% of the Big-Three average. Long-term debt/total liabilities for D-B for 1993-1995 more or less mirror the figures for Volkswagen along with the same trend. Audi and Ford-Werke do not have long-term debt. The 1996 figure is 60% of the Big-Three average and 86% of Chrysler's figure. For 1997, D-B's long-term debt/total liabilities is 71% of the Big-three average and 92% of Chrysler's figure. Like its German peers, D-B has made limited use of financial leverage to finance its operations. Although its weaker operating performance has provided lower coverage for debt service, its use of financial leverage is similar to Chrysler's.

A traditional measure of the market risk of the common equity of a public company is beta. Both Global Vantage and PC Plus include beta estimates for public companies with at least five years of equity market returns' data. Global Vantage reports beta estimates of 1.159, 0.862, 1.646, 0.049, and 1.020 for D-B, Audi, BMW, Ford-Werke, and Volkswagen, respectively. No beta is reported for Adam Opel, as General Motors owns 100% of its common equity. The Global Vantage beta estimates use the German DAX index to measure market returns. Using the S&P 500 equity market index, PC Plus reports betas of 0.9, 0.9, 0.7, and 0.9 for D-B, Chrysler, Ford, and General Motors. D-B's market risk appears to be in the range of 90% to 116% of the market index in recent years, about average for both German and U.S. automobile manufactures.

In summary, the ratio analysis suggests a mixed performance for D-B relative to its German peer companies from 1993 to 1995. D-B's financial performance generally improved from 1993 to 1994, while its performance in 1995 reflects a "big bath" with significant net losses. Compared to Chrysler's performance from 1996 to 1997, D-B's profitability and activity measures are weaker. It has higher measures of liquidity for 1996 and 1997. Also, its use of financial leverage is similar to Chrysler's. Thus Chrysler's financial performance in recent years appears to complement D-B's, and would appear to support the merger of these two companies.

CONCLUSION

This paper furnishes both a classroom guide on analysis of financial reports, especially in the international realm, and a comparative analysis of financial report databases to use for such a project. Faculty could benefit from this paper in assigning these projects and recommending particular databases for the university library to acquire.

From our experience, motivated students enjoy doing the financial report project, deriving considerable knowledge and insight from the overall assignment. About half the students requested and received individual attention or general guidance from us on their project. We provided individual assistance in selecting the topic or companies, recommended research sources, and reviewed first drafts of the paper. Students achieve a better understanding of U.S. and foreign GAAP in doing this project, judging from their oral feedback and course evaluations. Also, the students appreciate the linkage between the culture of the country and its accounting framework and principles. In making class presentations, the students enjoy sharing with their peers their own learning experiences about the culture and GAAP in different countries throughout the world. A few students have suggested group projects for those taking the course in the future in order to ease their workload. As the enrollment in this course expands, students will do at least one of the course projects in groups.

APPENDIX A

Department of Accountancy

SYLLABUS

International Accounting

Accounting 561 (M.B.A)

Spring 1999

PRINCIPAL TEXT: Lee H. Radebaugh and Sidney J. Gray, *International Accounting and Multinational Enterprise*, 4th ed., John Wiley, 1997.

COURSE DESCRIPTION: This is a graduate course intended to provide comprehensive coverage of the international aspects of accounting with particular emphasis on financial reporting and limited exposure to management control.

COURSE OBJECTIVES:

1. To achieve an understanding of differences in financial accounting principles from country to country and to attempt to understand such differences.
2. To better understand financial reports of foreign-based corporations. To consider issues in management control, budgeting, and performance evaluation of global operations.

ASSIGNMENTS, TESTS, AND GRADES:

1. Chapter readings, questions, problems, and cases will be assigned during each class session to be covered in the subsequent class session. *Your analysis and answers to each question, case, or assignment must be written and completed prior to each class session.* You are expected to be prepared in class to discuss the assigned reading material and to describe what you used to arrive at the answer to every question, problem, and case.

2. Twenty percent of your course grade is based on the quality of your assignments as perceived by your professor in class. The schedule of assignments indicated below is not fixed, and will be modified somewhat during the course as developments warrant.
3. Your course grade is determined as follows:

Assignments	20%
Cases	25%
Paper	120%
Paper	235%
	100%

Grade Distribution

90–100	A
80–89	B
70–79	C
60–69	D
Below 60	F

4. Plagiarism in this class, whether on a test, a case, or an assignment, will not be tolerated.

AGENDA

SESSION	DATE	TOPIC	CHAPTER
1	1/11	**Overview and Introduction**	1
		Cases: Social Financial Reporting *In India* BOC (United Kingdom)	
2	1/25	**International Business and Corporate Strategy**	2,3
		Case: Proctor and Gamble	
3	2/01	**Accounting Systems in Different Countries**	4,5
		In-class Case: Armenia	
4	2/08	**Accounting Systems in Different Countries**	6
		Accounting Harmonization	
		Case: International Harmonization	
5	2/15	**Disclosure and Regulation**	7
		Case: Volvo	
		PAPER #1 DUE: Presentation and Submission	
6	2/22	**Business Combinations**	8,9,10
		Goodwill and Segment Reporting	
		Cases: Smith Kline Beecham (U.K.), BASF (Germany)	
7	3/08	**Inflation Accounting**	
		In-class Case: Problems	
8	3/15	**Foreign Currency Translation**	13
		Cases: RadCo International (3,4)/Kamikaze/Daimler-Benz	
9	3/22	**Foreign Currency Transactions**	12
		Case: BP (U.K.)	

(continued)

AGENDA

SESSION	DATE	TOPIC	CHAPTER
10	3/29	**Comparative Analysis of Financial Statements**	14,15
		Cases: Japanese Puzzle/Matterhorn/Daimler-Benz	
11	4/06	**Global Management Control**	16,17
		Cases: Electrolux Autoparts/Star Oil	
12	4/12	**Global Performance Measurement**	17, 18
13	4/19	**Global Issues and PAPER #2 Presentations**	21
14	4/26	**PAPER #2 DUE: Presentations**	
15	5/03	**FINAL EXAMINATION**	

TERM PAPERS

The specific goals of each term paper are:

1. To stimulate critical thinking on the part of the student on the reasons for differences in accounting form country to country.
2. To give students an opportunity to prepare a report involving the analysis of accounting data from an international perspective.
3. To emphasize the nature of professional writing in terms of the preparation of various drafts, revisions, and editing.
4. To encourage students to exchange drafts of papers with each other for peer review and later with the instructor as well.

The papers will be graded on the basis of content, organization, and style. Emphasis will be placed on clarity and conciseness. The first paper should be no more than 10 pages, not including exhibits. The second paper should be at least 15 pages.

PAPER #1:

You are expected to select a country, to develop a conceptual framework for that country, and to prepare an analysis of the main generally accepted accounting principles (GAAP) and methods used in that country. You are to compare those GAAP with their U.S. counterparts, clearly discussing the similarities and differences between them. Finally, you are to attempt to explain those differences in terms of specific cultural, historical, economic, legal, and social aspects of the two countries.

PAPER #2:

Option 1: The objective of this assignment is for you to evaluate the performance of a particular company, primarily from an examination of its annual reports. In the process, you should examine the financial statements, footnotes, companion schedules, audit reports, and management discussion contained in the annual reports. You should also examine industry journals relevant to the company as well as newspaper articles about the company. Additionally, you are encouraged to contact the financial liaison officer at the company to secure answers to questions you have about the recent performance of the firm.

You are expected to analyze where the company has been in recent years, where it is heading—to map out an expected scenario for the firm—by synthesizing information from various sources about the company. Emphasis is placed on you perception of the firm, not just from the quantitative, financial data but also from qualitative information.

Option 2:   This option involves writing your own case and discussion notes. There are basically two types of cases:

1.   The closed system, where all the relevant data are included in the case.
2.   The open system, where all the relevant data are not included in the case and, therefore, students have to do library research, search databases, and conduct interviews.

A case may involve an annual report of a foreign company or may stem from a current events article on a particular foreign company, but should focus on issues in international accounting that are covered in this course. Each case should be in the same format:

Section I	Case
Section II	Exhibits
Section III	Discussion Notes

Option 3:   This option involves writing an analysis of how a particular accounting issue or concept is treated in a number of different countries throughout the world and attempting to explain the differences based on culture.

APPENDIX B

Peer Group Analysis Using Compustat Global Vantage and PC Plus Databases

This exercise initially assumes the user has a personal floppy disk on the A drive and is using the Global Vantage database. To obtain data for individual companies, you will need to know the Global Vantage Company Key (GVKEY) for each company, a unique identification number for each company in the Global Vantage database. The "Look Up" function of the database (the button with Eye Glasses) allows you to determine the GVKEY from the company name as well as all other accessible fields in the data base. The following steps will generate a peer group analysis using the Compustat Global Vantage database:

1.   Obtain the Compustat Global Vantage CD disk, and install the disk in the CD drive of the designated PC.
2.   Open Global Vantage. (Double click on the PC Plus icon.)

To determine a peer group set of companies for a particular SIC code and a particular country:

1. Click on Screen, click on New, and click on Formula cell. Type SIC, and hit Enter.
2. Type a numeric SIC code (autos = 3711) in Min cell, and hit Enter. Type the same numeric SIC cod in Max cell, and hit Enter. Again hit Enter to move next row of the Screen window.
3. Type CINC in second Formula cell and hit Enter. Type a country id, and hit Enter for both Min and Max cells. [Country ids can be found under Help/Country of Incorporation Codes In Alphabetical Order (Germany = deu)].
4. Run the screen using the Run icon (a Runner), or right click on CINC. Then click run.
5. After the Run is complete, right click on CINC again, and click Save to save your set of companies to a file on your disk in the A drive, using a user-defined file name.

Having identified the particular company and the peer group set of companies with the screen procedure, you can now apply the preprogrammed reports included in the database. Useful reports include Comparative Ratio Report (includes same ratios of the previous report for up to six user defined firms for a single year), Annual Ratio—Five Years (ratios for a single company for each of five years for a user-defined, five-year period), and Performance Summary (a ten-year profitability ratio and growth rate report for a single company).

You should follow these commands to access the Comparative Ratio Report:

1. Click on Set button. Go to Drive A, and click on the file name of the saved screen set.
2. Click on the Report button. Then click on Annual Ratio—5 Years. Click OK. Note that you can begin by using the current period as the base period, or you can choose a different period for the base period by utilizing the Select Period option.
3. Click on the Set File button, and choose your file of companies from drive A. Then click OK.
4. Click OK on the Run Report window. When the run is completed, you will be able to see the report for the first company listed in your set. By clicking the blue arrow buttons on the task bar, you can view separate reports for each company in your company set.
5. To print these reports click on the Printer button. You can choose either the current company on your screen or all companies in your identified set. Make this choice, and click OK. Note that these reports can be saved to your disk as a text file for subsequent printing, using the save file button. You can save a report for either the current company on your screen or all companies in the identified set.

To utilize the Comparative Ratio Report or the Performance Report, repeat the previous procedure beginning with Step 2 above. Currently, Global Vantage does not provide a report that computes average ratio values for a set of firms. You can develop such a report using the Report creation features of the software, although more advanced knowledge of this feature requires reading the appropriate sections in the PC Plus Basics manual. Alternatively, you can save the previously reported data to file and then process with spreadsheet software to calculate average ratio values for a particular set of firms. The same procedures can be utilized with PC Plus to generate the reports described above, except that individual companies are uniquely identified by their ticker symbol.

Table B1.   Comparative Annual Ratio Report—1993
(Compustat Global Vantage)

COMPANY A: DAIMLER-BENZ AG	COMPANY D: BMW AG
COMPANY B: ADAM OPEL AG	COMPANY E: FORD WERKE AG
COMPANY C: AUDI AG	COMPANY F: VOLKSWAGEN AG

COMPANY	A	B	C	D	E	F	AVERAGE
	Dec-93	Dec-93	Dec-93	Dec-93	Dec-93	Dec-93	VALUE
ISO CODE:	DEM	DEM	DEM	DEM	DEM	DEM	B,C,D,E,F
LIQUIDITY							
Current Ratio	2.37	2.05	3.77	1.50	1.64	1.81	2.15
Quick Ratio	1.45	1.30	2.85	0.85	1.15	0.72	1.37
ACTIVITY							
Inventory Turnover	0.00	0.00	16.39	0.00	0.31	6.11	4.56
Receivables Turnover	4.34	7.68	18.9	13.61	8.50	14.16	10.57
Total Asset Turnover	1.10	2.13	2.12	1.00	2.47	0.99	1.74
Average Collection Period (Days)	84.13	47.55	19.30	101.10	42.93	25.78	47.33
Days to Sell Inventory	NC	NC	22.28	NC	1196.48	49.47	422.74
Operating Cycle (Days)	NC	NC	41.58	NC	1239.41	85.54	455.51
PERFORMANCE							
Revenues/Net Fixed Assets	3.17	6.93	4.76	4.31	5.68	2.50	4.84
Revenues/Stockholders Equity	5.56	13.49	8.55	4.20	30.13	6.80	12.63
PROFITABILITY							
Operating Margin Before Depr (%)	6.87	2.16	NC	9.13	5.55	3.63	5.12
Operating Margin After Depr (%)	−1.37	−2.77	−2.81	2.81	−0.43	−2.65	−1.17
Pretax Profit Margin (%)	1.56	−1.48	−1.18	2.87	0.00	−2.14	−0.39
Net Profit Margin (%)	0.62	−2.18	−0.71	1.81	-0.62	−2.66	−0.87
After Tax Profit Margin—Including							
Extraordinary Items (%)	0.62	−2.18	−0.71	1.81	−0.62	−2.66	-0.87
Return on Assets (%)	0.68	−4.64	−1.50	1.81	−1.54	−2.64	−1.70
Return on Common Equity (%)	3.34	−24.40	−5.89	7.80	−17.16	−16.98	−11.33

(continued)

Table B1 (Continued)

COMPANY	A	B	C	D	E	F	AVERAGE
	Dec-93	Dec-93	Dec-93	Dec-93	Dec-93	Dec-93	VALUE
ISO CODE:	DEM	DEM	DEM	DEM	DEM	DEM	B,C,D,E,F
Return on Total Equity (%)	3.34	−24.40	−5.89	7.74	−17.16	−16.44	−11.23
Return on Investment (%)	2.07	−28.73	−5.96	7.46	−18.77	−8.37	−10.87
Total Expense/Total Revenue (%)	NA	NA	NA	NA	NA	NA	
LEVERAGE							
Interest Coverage Before Tax	1.75	−6.17	−18.20	1.82	1.00	0.35	−4.24
Interest Coverage After Tax	1.30	−9.60	−10.56	1.51	0.42	0.22	−3.60
Long-Term Debt/Common Equity (%)	62.50	2.19	0.00	0.00	0.00	95.79	19.60
Long-Term Debt/Shrhldr Equity (%)	62.50	2.19	0.00	0.00	0.00	93.02	19.04
Total Debt/Invested Capital (%)	67.35	2.57	0.00	153.44	0.00	105.50	52.30
Total Debt/Total Assets (%)	21.58	0.44	0.00	35.58	0.00	32.41	13.69
Total Assets/Common Equity	5.17	5.94	4.14	4.42	11.74	7.25	6.70
Long-Term Financing Ratio	30.42	19.10	25.52	32.15	8.94	28.53	22.85
Capitalization Ratio	20.40	18.89	25.52	23.46	8.94	16.09	18.58
Long-Term Borrowings/Total Liabs (%)	14.98	0.44	0.00	0.00	0.00	15.40	3.17

Notes: NA = Not available
NC = Not calculable

Table B2. Comparative Annual Ratio Report—1994 (Compustat Global Vantage)

COMPANY A: DAIMLER-BENZ AG	COMPANY D: BMW AG
COMPANY B: ADAM OPEL AG	COMPANY E: FORD WERKE AG
COMPANY C: AUDI AG	COMPANY F: VOLKSWAGEN AG

COMPANY	A	B	C	D	E	F	AVERAGE
	Dec-94	Dec-94	Dec-94	Dec-94	Dec-94	Dec-94	VALUE
ISO CODE:	DEM	DEM	DEM	DEM	DEM	DEM	B,C,D,E,F
LIQUIDITY							
Current Ratio	2.22	2.15	2.92	1.94	1.93	1.95	2.18
Quick Ratio	1.48	1.38	2.29	1.01	1.41	0.86	1.39
ACTIVITY							
Inventory Turnover	4.65	0.00	17.89	0.00	0.33	6.22	4.89
Receivables Turnover	4.26	11.60	45.18	4.67	8.26	13.23	16.59
Total Asset Turnover	1.13	2.41	2.10	1.22	2.75	1.00	1.90
Average Collection Period (Days)	85.63	31.46	8.08	78.10	44.20	27.60	37.89
Days to Sell Inventory	78.49	NC	20.40	NC	1112.65	46.22	393.09
Operating Cycle (Days)	164.12	NC	28.48	NC	1156.85	86.25	423.86
PERFORMANCE							
Revenues/Net Fixed Assets	3.73	8.86	4.53	3.85	6.81	2.79	5.37
Revenues/Stockholders Equity	5.18	14.02	9.00	5.40	16.91	7.29	10.52
PROFITABILITY							
Operating Margin Before Depr (%)	8.91	3.89	CF	9.52	7.27	8.36	7.26
Operating Margin After Depr (%)	1.78	−0.32	0.19	3.42	2.35	−0.19	1.09
Pretax Profit Margin (%)	2.00	1.16	1.45	3.22	3.80	0.58	2.04

(*continued*)

Table B2   (Continued)

COMPANY	A	B	C	D	E	F	AVERAGE
	Dec-94	Dec-94	Dec-94	Dec-94	Dec-94	Dec-94	VALUE
ISO CODE:	DEM	DEM	DEM	DEM	DEM	DEM	B,C,D,E,F
Net Profit Margin (%)	0.71	1.20	0.22	1.65	2.89	0.19	1.23
After Tax Profit Margin—Including		1.20	0.22	1.65	2.89	0.19	1.23
Extraordinary Items (%)	0.71						
Return on Assets (%)	0.80	2.89	0.46	2.01	7.93	0.19	2.70
Return on Common Equity (%)	3.91	17.39	2.00	9.24	64.81	1.14	18.92
Return on Total Equity (%)	3.91	17.39	2.00	9.42	64.81	1.34	18.99
Return on Investment (%)	2.44	15.47	1.95	4.81	48.87	0.64	14.35
Total Expense/Total Revenue (%)	NA	NA	NA	NA	NA	NA	
LEVERAGE							
Interest Coverage Before Tax	2.43	5.89	19.61	2.04	11.43	1.21	8.04
Interest Coverage After Tax	1.62	6.04	3.82	1.54	8.93	1.07	4.28
Long-Term Debt/Common Equity (%)	49.58	0.47	0.00	83.90	0.00	97.91	36.46
Long-Term Debt/Shrhldr Equity (%)	49.58	0.47	0.00	83.21	0.00	94.98	35.73
Total Debt/Invested Capital (%)	65.14	0.50	0.00	97.50	1.58	105.20	40.96
Total Debt/Total Assets (%)	21.04	0.09	0.00	36.32	0.25	30.38	13.41
Total Assets/Common Equity	4.65	6.11	4.52	5.00	6.35	7.61	5.92
Long-Term Financing Ratio	31.78	16.83	23.14	29.88	12.13	26.92	21.78
Capitalization Ratio	20.41	16.60	23.14	21.49	12.13	13.88	17.45
Long-Term Borrowings/Total Liabs (%)	13.57	0.09	0.00	21.01	0.00	14.89	7.20

Notes: NA = Not available
NC = Not calculable

Table B3.   Comparative Annual Ratio Report—1995
(Compustat Global Vantage)

COMPANY A: DAIMLER-BENZ AG	COMPANY D: BMW AG
COMPANY B: ADAM OPEL AG	COMPANY E: FORD WERKE AG
COMPANY C: AUDI AG	COMPANY F: VOLKSWAGEN AG

COMPANY	A	B	C	D	E	F	AVERAGE
	Dec-95	Dec-95	Dec-95	Dec-95	Dec-95	Dec-95	VALUE
ISO CODE:	DEM	DEM	DEM	DEM	DEM	DEM	B,C,D,E,F
LIQUIDITY							
Current Ratio	2.43	2.09	3.15	2.03	1.33	1.95	2.11
Quick Ratio	1.37	1.37	2.55	1.03	0.84	0.86	1.33
ACTIVITY							
Inventor Turnover	5.92	0.00	20.6	50.00	0.48	7.62	5.75
Receivable Turnover	3.93	12.17	67.03	4.63	9.80	12.88	21.30
Total Asset Turnover	1.05	2.33	2.20	1.16	2.75	1.07	1.90
Average Collection Period (Days)	92.87	29.98	5.45	78.79	37.24	28.34	35.96
Days to Sell Inventory	61.70	NC	17.67	NC	761.70	38.60	272.66
Operating Cycle (Days)	154.57	NC	23.12	NC	798.94	76.21	299.42

(continued)

Table B3   (Continued)

COMPANY	A	B	C	D	E	F	AVERAGE
	Dec-95	Dec-95	Dec-95	Dec-95	Dec-95	Dec-95	VALUE
ISO CODE:	DEM	DEM	DEM	DEM	DEM	DEM	B,C,D,E,F
PERFORMANCE							
Revenues/Net Fixed Assets	3.94	9.42	5.05	4.16	7.29	3.08	5.80
Revenues/Stockholders Equity	4.51	13.71	9.45	5.72	15.85	8.38	10.62
PROFITABILITY							
Operating Margin Before Depr (%)	NC	5.68	NC	9.67	5.40	7.22	6.99
Operating Margin After Depr (%)	−7.93	1.45	2.26	3.43	0.96	0.13	1.65
Pretax Profit Margin (%)	−7.02	1.55	3.53	2.96	2.04	1.26	2.27
Net Profit Margin (%)	−5.56	1.40	0.67	1.49	1.08	0.40	1.01
After Tax Profit Margin—Including							
Extraordinary Items (%)	−5.56	1.40	0.67	1.49	1.08	0.40	1.01
Return on Assets (%)	−5.86	3.27	1.47	1.73	2.97	0.43	1.97
Return on Common Equity (%)	−26.67	19.55	6.80	8.49	18.26	2.96	11.21
Return on Total Equity (%)	−26.67	19.55	6.80	8.66	18.26	3.29	11.31
Return on Investment (%)	−17.44	17.76	6.21	4.36	17.13	1.62	9.42
Total Expense/Total Revenue (%)	NA	NA	NA	NA	NA	NA	
LEVERAGE							
Interest Coverage Before Tax	−6.18	12.42	39.17	1.84	6.79	1.47	12.34
Interest Coverage After Tax	−4.57	11.35	8.20	1.42	4.07	1.14	5.24
Long-Term Debt/Common Equity (%)	37.93	0.33	0.00	94.24	0.00	99.11	38.74
Long-Term Debt/Shrhldr Equity (%)	37.93	0.33	0.00	93.46	0.00	95.98	37.95
Total Debt/Invested Capital (%)	67.83	0.31	0.00	92.30	0.00	106.55	39.83
Total Debt/Total Assets (%)	21.83	0.06	0.00	35.58	0.00	27.78	12.68
Total Assets/Common Equity	4.47	5.86	4.74	5.10	5.96	8.25	5.98
Long-Term Financing Ratio	31.51	16.78	21.62	37.59	16.25	25.47	23.54
Capitalization Ratio	21.94	16.71	21.62	19.96	16.25	13.03	17.51
Long-Term Borrowings/Total Liabs (%)	10.94	0.07	0.00	23.03	0.00	13.72	7.36

Notes: NA = Not available
NC = Not calculable

Table B4.   Comparative Annual Ratio Report—1996 (Compustat PC Plus)

COMPANY A: DAIMLER-BENZ SPN	COMPANY C: FORD				
COMPANY B: CHRYSLER	COMPANY D: GENERAL MTRS				
	A	B	C	D	AVERAGE
COMPANY	Dec-96	Dec-96	Dec-96	Dec-96	B,C,D
LIQUIDITY					
Current Ratio	1.55	1.11	1.69	1.08	1.29
Quick Ratio	1.09	0.82	1.58	0.94	1.11
Working Capital Per Share	26.97	3.83	67.98	9.04	26.95
Cash Flow Per Share	10.98	8.59	14.50	21.98	15.02
ACTIVITY					
Inventory Turnover	4.81	9.76	17.07	10.81	12.55
Receivables Turnover	3.14	3.94	0.92	2.41	2.42

(continued)

Table B4 (Continued)

	A	B	C	D	AVERAGE
COMPANY	Dec-96	Dec-96	Dec-96	Dec-96	B,C,D
Total Asset Turnover	0.96	1.08	0.58	0.73	0.80
Average Collection Per (Days)	115.00	91.00	391.00	149.00	210.33
Days to Sell Inventory	75.00	37.00	21.00	33.00	30.33
Operating Cycle (Days)	190.00	128.00	412.00	183.00	241.00
PERFORMANCE					
Sales/Net PP&E	3.53	3.15	4.38	2.37	3.30
Sales/Stockholder Equity	4.03	5.13	5.49	6.84	5.82
PROFITABILITY					
Oper.Margin Before Depr (%)	6.57	12.70	19.77	12.60	15.02
Oper.Margin After Depr (%)	0.06	8.80	11.06	5.30	8.39
Pretax Profit Margin (%)	1.84	10.27	4.62	4.17	6.35
Net Profit Margin (%)	2.60	6.27	3.02	3.09	4.13
Return on Assets (%)	2.46	6.62	1.67	2.19	3.49
Return on Equity (%)	10.47	32.12	16.37	20.81	23.10
Return on Investment (%)	7.04	19.82	4.19	7.84	10.62
Return on Average Assets (%)	2.49	6.76	1.73	2.22	3.57
Return on Average Equity (%)	13.20	33.00	17.08	20.84	23.64
Return on Average Invest. (%)	8.50	18.79	4.30	7.94	10.34
LEVERAGE					
Interest Coverage Before Tax	3.13	6.24	1.65	2.17	3.35
Interest Coverage After Tax	4.00	4.20	1.43	1.87	2.50
Long-Term Debt/Common Eq. (%)	45.03	62.09	290.77	165.26	172.71
Long-Term Debt/Shrhldr Eq. (%)	45.03	62.09	290.77	165.25	172.70
Total Debt/Invested Cap. (%)	69.32	71.43	151.94	138.33	120.57
Total Debt/Total Assets (%)	24.17	23.84	60.50	38.68	41.01
Total Assets/Common Equity	4.26	4.86	9.82	9.49	8.06
Long-term Debt/Total Liabilities	13.81	16.10	32.96	19.47	22.84
DIVIDENDS					
Dividend Payout (%)	0.00	27.33	39.60	31.40	32.78
Dividend Yield (%)	0.42	4.24	4.56	2.87	3.89

Table B5. Comparative Annual Ratio Report—1997 (Compustat PC Plus)

| COMPANY A: DAIMLER-BENZ SPN | | COMPANY C: FORD | | | |
| COMPANY B: CHRYSLER | | COMPANY D: GENERAL MTRS | | | |

	A	B	C	D	AVERAGE
COMPANY	Dec-97	Dec-97	Dec-97	Dec-97	B,C,D
LIQUIDITY					
Current Ratio	1.60	1.05	1.72	0.98	1.25
Quick Ratio	1.17	0.78	1.62	0.85	1.08
Working Capital Per Share	34.16	1.98	74.96	−2.20	24.91
Cash Flow Per Share	16.74	8.48	17.04	33.27	19.60
ACTIVITY					
Inventory Turnover	6.37	9.56	19.73	11.67	13.65
Receivables Turnover	3.25	3.96	0.89	2.58	2.48
Total Asset Turnover	0.97	1.01	0.57	0.75	0.78
Average Collection Per (Days)	111.00	91.00	405.00	140.00	212.00
Days to Sell Inventory	56.00	38.00	18.00	31.00	29.00
Operating Cycle (Days)	167.00	129.00	424.00	170.00	241.00
PERFORMANCE					
Sales/Net PP&E	3.49	2.60	4.44	2.48	3.17
Sales/Stockholder Equity	3.54	5.16	5.00	9.61	6.59
PROFITABILITY					
Oper.Margin Before Depr (%)	7.68	10.57	22.13	16.73	16.48
Oper.Margin After Depr (%)	1.62	5.97	13.29	6.99	8.75
Pretax Profit Margin (%)	3.43	7.77	7.12	4.59	6.49
Net Profit Margin (%)	6.48	4.78	4.50	3.98	4.42
Return on Assets (%)	5.87	4.64	2.46	2.88	3.33
Return on Equity (%)	22.92	24.68	22.34	37.70	28.24
Return on Investment (%)	15.04	13.77	6.14	10.82	10.24
Return on Average Assets (%)	6.30	4.81	2.53	2.93	3.42
Return on Average Equity (%)	25.60	24.45	23.88	32.26	26.86
Return on Average Invest. (%)	16.99	14.33	6.35	10.72	10.47
LEVERAGE					
Interest Coverage Before Tax	5.09	4.80	2.04	2.26	3.03
Interest Coverage After Tax	8.73	3.34	1.66	2.10	2.37
Long-Term Debt/Common Eq. (%)	49.07	79.26	263.30	244.43	195.66
Long-Term Debt/Shrhldr Eq. (%)	49.07	79.26	263.30	244.41	195.66
Total Debt/Invested Cap. (%)	73.50	76.03	151.16	153.79	126.99
Total Debt/Total Assets (%)	28.67	25.63	60.53	41.00	42.39
Total Assets/Common Equity	3.91	5.32	9.08	13.08	9.16
Long-term Debt/Total Liabilities	16.88	18.36	32.58	20.24	23.73
DIVIDENDS					
Dividend Payout (%)	0.00	38.37	28.63	24.55	30.52
Dividend Yield (%)	0.89	4.55	3.39	3.29	3.74

APPENDIX C

Peer Group Analysis Using Moody's International Company Data

The following steps generate a peer group analysis using the most recent year of financial statement data in the database and the default format of the Moody's database:

1. Open Moody's International Database.
2. Search by company name to obtain your individual company files in order to determine primary SIC code. Type the company name, and hit Enter. Double-click on company name to obtain the company file. The primary SIC code will be at the top of the company file. You can perform peer group analysis once you have defined a peer group using the primary SIC code.
3. Click on the search button. Scroll down to Primary SIC Code (or Any SIC Code), and click. Check Equal To, and enter the SIC number, and click OK. Wait for search to be completed, and choose the company of interest.
4. Choose the peer group folder for peer group analysis. If the peer group folder is not active, follow the help menu for peer group to create the folder.

Table C1.   Moody's Company Data Report
Company: Daimler-Benz A.G. (Germany, Fed. Rep.)

Country: Germany—(7 in group of 82)
Region: Europe—(32 in group of 82)

Peer Group	Company	Group	Country	Region
Return on equity	10.46	7.94	12.0	12.58
Return on assets	2.46	3.07	2.56	3.58
Profit margin	2.56	3.97	1.67	5.79
Current ratio	1.91	3.93	31.63	7.93
Working capital/Total assets	0.28	0.11	0.42	0.19
Current liabilities/Equity	1.33	1.70	0.74	2.03
Long term debt/Equity	0.00	0.52	0.09	0.58
Total liabilities/Equity	3.26	2.61	4.00	3.24

Note: © 1998 by Moody's Investors Service, Inc. All Rights Reserved.

APPENDIX D

Peer Group Analysis Using Disclosure/worldscope Database

The following steps generate a peer group analysis using Disclosure/Worldscope database:

1. Access the Disclosure WEB site (www.disclosure.com), and click on "Use Global Access."
2. Provide your username and password to open the database.
3. Identify either "Company Name," "Ticker Symbol," or "CUSIP" to bring up company files and reports.
4. Under "Database Reports," left click on "Fundamental Ratios" to display 29 fundamental ratios for the identified company for the most recent three years of financial statement data.
5. Access annual financial statement data under "Database Reports," and save to file using the WEB browser. Repeating this procedure for a defined set of peer group firms allows the construction of specific ratios for each firm for the past five years as well as average ratios values for all peer group firms using spreadsheet software.

Table D1. Daimler Benz AG
Fundamental Ratios

Fiscal Year Ending	12/31/97	12/31/96	12/31/95
Quick Ratio	0.97	0.95	0.92
Current Ratio	1.41	1.43	1.52
Sales/Cash	6.05	7.42	8.46
SG&A/Sales	0.14	0.15	0.20
Receivables Turnover	3.27	3.55	4.14
Receivables Days Sales	110.07	101.28	86.90
Inventories Turnover	8.62	7.82	7.19
Inventories Days Sales	41.76	46.05	50.09
Net Sales/Working Capital	5.04	5.28	4.97
Net Sales/Plant & Equipment	3.49	3.53	3.92
Net Sales/Current assets	1.46	1.59	1.69
Net Sales/Total Assets	0.90	0.95	1.01
Net Sales/Employees	413,406	366,650	331,149
Total Liab/Total Assets	0.74	0.76	0.76
Total Liab/Invested Capital	2.87	3.23	3.41
Total Liab/Common Equity	2.97	3.34	3.62
Times Interest Earned	NA	NA	NA
Current Debt/Equity	NA	NA	NA
Long Term Debt/Equity	NA	NA	NA
Total Debt/Equity	NA	NA	NA
Total Assets/Equity	3.91	4.26	4.47
Pretax Inc/Net Sales	0.03	0.02	−0.07
Pretax Inc/Total Assets	0.03	0.02	−0.07
Pretax Inc/Invested Capital	0.12	0.07	−0.32
Pretax Inc/Common Equity	0.13	0.08	−0.34
Net Income/Net Sales	0.06	0.03	−0.06
Net Income/Total Assets	0.06	0.02	−0.06
Net Income/Invested Capital	0.23	0.10	−0.25
Net Income/Common Equity	0.24	0.11	−0.27

Note: Copyright 1998 Disclosure Inc. All Rights Reserved

ACKNOWLEDGMENT

Standard & Poor's COMPUSTAT, Englewood, CO grants us permission to use the exhibits on pp. 154-159, Moody's Investor Service, New York, NY grants us permission to use the exhibit on p. 160, and Disclosure Inc., Bethesda, MD grants us permission to use the exhibit on p. 161.

NOTES

1. General Motors acquired Adam Opel in 1929. It produces European models for GM as well as the Cadillac Catera, which is exported to the U.S. (Moody's International Company Data Report 1998).
2. Ford-Werke is a German unit of Ford (U.S.), producing Ford products geared to the German and European markets, including passenger cars, commercial vehicles and tractors and parts.

REFERENCES

Accounting Education Change Commission. 1990. *Position Statement No. 1*, Objectives of education for accountants.

Alexander, D., and S. Archer, eds. 1992. *The European Accounting Guide*. London: Academic Press.

American Accounting Association Committee on the Future, Structure, Content, and Scope of Accounting Education. (Bedford Report) 1986. Future accounting education: Preparing for the expanding profession. *Issues in Accounting Education* (Spring): 168-195.

Avila, J. 1998. *NBC Nightly News*, November 12.

Center for International Financial Analysis and Research. 1993. *The Global Company Handbook*. Princeton, NJ: CIFAR.

Choi, F.D.S., C.A. Frost, and G. Meek. 1999. *International Accounting*, 3rd ed. Upper Saddle River, NJ: Prentice-Hall.

Disclosure. 1999. *Disclosure Global Access*. Bethesda, MD: Primark.

Dun & Bradstreet. 1999. *Industry Norms and Key Business Ratios*. New York: Dun & Bradstreet.

Gray, S.J. 1988. Towards a theory of cultural influences on the development of accounting systems internationally. *Abacus* (March): 1-15.

Hofstede, G. 1980. *Culture's Consequences: International Differences in Work-related Values*. Beverly Hills, CA: Sage.

Holt, P., and C. Hein. 1999. *International Accounting*, 4th ed. Houston, TX: Dame.

Iqbal, M., T. Melcher, and A. Elmallah. 1997. *International Accounting*. Cincinnati, OH: South-Western.

Kantor, J., C.B. Roberts, and S.B. Salter. 1995. Financial reporting in selected arab countries. *International Studies of Management & Organization* 25(3): 31-50.

Lenox, C. 1998. Success factors for supporting compustat in academic business libraries. *BFBulletin* (Fall): 41-49.

Moody's. 1998. *Moody's International Company Data*. New York: Moody's.

Mueller, G., H. Gernon, and G. Meek. 1997. *Accounting: An International Perspective*, 4th ed. Boston: Irwin McGraw-Hill.

O'Connor, W., D. Rapaccioli, and P. Williams. 1996. Internationalizing the advanced accounting course. *Issues in Accounting Education* (Fall): 315-335.

Radebaugh, L. and S. Gray. 1997. *International Accounting and Multinational Enterprises*, 4th ed. New York: Wiley.

Robert Morris Associates. 1998. *Annual Statement Studies.* Philadelphia, PA: Robert Morris Associates.

Standard & Poor's. 1998. *Compustat PC Plus.* Englewood, CO: Standard & Poor's.

Standard & Poor's. 1998. *Compustat Global Vantage.* Englewood, CO: Standard & Poor's.

Stout, D., and J. Schweikart. 1989. The relevance of international accounting to the accounting curriculum: A comparison of practitioner and educator opinion. *Issues in Accounting Education* (Fall): 126-143.

Tondkar, R., A. Adhikari, and E. Coffman. 1994. Adding an international dimension to upper-level financial accounting courses by utilizing foreign annual reports. *Issues in Accounting Education* (Fall): 271-281.

Troy, L. 1999. *Almanac of Business and Industrial Ratios.* Englewood Cliffs, NJ: Prentice-Hall.

Wallace, W. 1996. Ratios for your use in extending internationalization in accounting courses. *Issues in Accounting Education* (Fall): 337-343.

THE EFFECT OF GROUP REWARDS ON OBTAINING HIGHER ACHIEVEMENT FROM COOPERATIVE LEARNING

Gary Grudnitski

ABSTRACT

This paper provides a replication of the work of Ravenscroft et al. (1997). Its purpose is to assess the impact on individual achievement in accounting courses when students engage in cooperative-learning activities and are rewarded, in part, based on the performance of all members of their group. Achievement was measured in a semester-long, introductory management accounting course by the difference between the postcooperative learning exam scores and the precooperative learning test scores for 100 students. This study finds that, after participating in cooperative-learning activities, students who were given bonuses based on the performance of their group on exams achieved a statistically higher mean difference than did students who were given bonuses based only on their performance on exams. Moreover, the achievement level of students offered bonuses based on the performance of their group on exams was approximately 4% higher ($p = 0.11$) than the achievement level of students offered bonuses based only on their individual exam performance.

Advances in Accounting Education, Volume 2, pages 165-177.
Copyright © 2000 by JAI Press Inc.
All rights of reproduction in any form reserved.
ISBN: 0-7623-0515-0

INTRODUCTION

The objective of this research is to replicate Ravenscroft et al. (1997) by determining the effect on individual achievement when students in an accounting course engage in cooperative-learning activities and are provided group versus individual incentives. This study's findings are important because they hold the potential of contributing to an accounting education knowledge base in the topical areas of the impact of cooperative-learning techniques on student performance and the influence of different approaches to group grading. Both of these areas of further study were suggested by the research agenda for accounting education presented by Stout and Rebele (1996, 10-11).

Toward accomplishing this objective, the research background for this study is provided first. Next, the results are reported for the achievement level attained in an introductory management accounting course when students are given bonuses, in part, based on the performance of all members of their group versus on their own performance only. Finally, in the last part of the paper, a scenario is offered to explain why achievement levels may vary within a cooperative-learning environment employing group versus individual reward structures.

PRIOR RESEARCH

The appeal for many accounting instructors to try some form of cooperative learning is based on research indicating that cooperative learning promotes higher achievement relative to competitive or individualistic learning (Johnson and Johnson 1994; Slavin 1995). However, simply putting students in groups and telling them to work together does not, in and of itself, promote higher achievement (Johnson and Johnson 1990). These authors (1990, 28) maintain that positive interdependence is the most important factor in structuring learning situations cooperatively, and Slavin (1990) adds that the most successful method for increasing student achievement is to provide group rewards that reflect the performance of all members of the group. That is, student achievement may be maximized in cooperative-learning settings when individual group members are rewarded, at least in part, on the collective performance of members in the group. Under the condition of group performance affecting individual rewards, the group will be more likely to focus on the task of ensuring that every member in the group has learned the subject matter.

In the accounting education literature, three research studies bear directly on the issue of the relationship between rewards for group-based performance and student achievement in cooperative-learning environments. In the first study, Ravenscroft et al. (1995) compare the effects on exam performance of individual and group grading for students in two sections of an accounting principles course. They report that the students who were graded on both their individual and their group's performance scored higher on exams and quizzes than the students who were graded entirely on their individual performance.

In a second study, Ciccotello et al. (1997) evaluate the in-class examination performance of undergraduate managerial accounting students in response to group problem-solving workshops. Although their results are consistent with Ravenscroft et al. (1995) in demonstrating how positive group interdependence can influence individual performance, the authors add the caveat (p. 6) that "...this study does not examine the impact of group grading per se."

By far the most extensive research into assessing group grade incentives on exam performance of individuals in accounting courses is the recent study by Ravenscroft et al. (1997). In brief, Ravenscroft et al. find little or no improvement on exam performance across a variety of accounting courses when group incentives are used or when a student team learning approach is adopted. They conclude that employing cooperative-learning activities and group incentives in accounting courses does not automatically produce positive achievement effects. While Ravenscroft et al.'s (1997) study offers the most comprehensive test to-date of the influence of cooperative-learning activities and group incentives on achievement in accounting courses, the manner in which their group grade incentive was operationalized may have biased against them finding statistically significant differences in achievement between their experimental and control sections. Specifically, in their experimental section exam scores were based on the following formula (p. 156):

$$\text{Individual score } (.70) + \text{Group average } (.30)$$

One can speculate on the disincentives this formula might have for high- and low-achieving students in a group. For the high-achieving students in a group, this group-grading formula may be regarded as a burden. These students may feel that despite their best efforts at getting everyone in the group "up to speed" on the material, the low-achieving students in the group will never equal the exam performance of the high-achieving students. For the group's low-achieving students, this group-grading formula may be regarded as a bounty because their scores will now reflect, in part, the group's high-achieving students' exam performance.

This research replicates the study of Ravenscroft et al. (1997) by asking the same question: Within a cooperative-learning environment, do students who are rewarded, in part, based on the individual exam performance of members of their group achieve more than other students who are rewarded solely on the basis of their own exam performance? This research serves as a differentiated replication (Lindsey and Ehrenberg 1993) of Ravenscroft et al.'s (1997) study by adopting an incentive system wherein students of all achievement levels can benefit. In general terms, this study's incentive system awards bonus points to all students in a group whenever any student in that group improves on his or her achievement score. Because the awarding of bonus points now reflects the *relative* improvement of a student after cooperative-learning activities have taken place, both high- and low-achieving students have proper incentives for cooperative learning. Specifically, the high-achieving student in a group has an incentive to help the low-achieving students learn because bonus points will be added to the exam score of the high-achieving student when the low achieving students in the group perform *relatively* better on their exams than on their individual mini-tests. Further, because the low-achieving students have the greatest room for improvement, the low-achieving students now realize that they can make the greatest difference in the bonus points received by their group.

METHOD

The setting for this study was two sections of a Managerial Accounting Fundamentals course. Managerial Accounting Fundamentals is the second accounting course in a twelve-course core sequence required of all proposed undergraduate business majors at a large U.S. university. Typically, Managerial Accounting Fundamentals is taken by students when they are at the sophomore level. Both sections of this semester-long class were taught by the same instructor, and used the same text, problem assignments, and testing material.

One section of the course had 54 students in it; the second section had 46 students. In the section with 54 students, students obtained an average of 2.75, on a 4.0 scale, in the prerequisite Introductory Financial Accounting course and had an overall grade point average of 2.81. Of the students in this section, 37% were females and 13% intended to major in accounting.

In the 46-student section, the students averaged 2.68 in the Introductory Financial Accounting course and had an overall grade point average of 2.80. Of the students in this section, 35% were females and 11% intended to major in accounting.

In each of the two sections, students were assigned randomly to a group of four to six persons at the second class meeting. (Variation in group size was caused by the normal amount of adding and dropping during the first three weeks of the semester as students adjusted their schedules.)

Sequence of Course Activities and Grade Components

The following (identical) sequence of course activities were experienced by students in both sections:

1. For each unit of material assigned and lectured on (typically, one text chapter), each student individually completed and turned in homework exercises from the text. Ten percent of a student's grade was based on completion of these homework assignments. Each homework assignment was scored on a pass/fail basis reflecting the effort exhibited in working the exercise rather than correctness of the answer to the exercise.

2.   As in the informative testing system advocated by Michaelsen et al. (1993), each student took a mini-test individually and as a member of that student's group. These mini-tests consisted of approximately 20 multiple-choice (MC) questions. Twenty percent of the course grade came from a student's individual performance on twelve mini-tests and 20% of the course grade reflected the performance of the student's group in retaking the mini-tests. Following taking a mini-test individually and in the same class period the student's group met and decided on the "best" answer to each mini-test question. Accordingly, all members of the group received the same score based on the correctness of the group's answers.

3.   Students worked as a group on a comprehensive problem and submitted a group solution. Ten percent of the course grade was determined by the correctness of solutions prepared by the group for twelve comprehensive problems.

4.   In the fifth, tenth, and fifteenth weeks of the semester, students took an individual comprehensive exam, which consisted of approximately 30 MC questions, on the material covered during that portion of the course. Each of the three individual exams contributed 10% toward a student's final grade.

5.   At the end of the semester, each student prepared a peer evaluation of the contribution made by all other members of the student's group to the quality of the group's work. This evaluation asked about the level of effort expended by each student in the group. It was primarily meant to encourage participation (discourage "free riding") by individual members of the group in the group activities, and comprised 10% of each student's final grade.

In summary of the above sequence of course activities, Table 1 lists the factors that went into determining a student's grade in the class and the points assigned to each factor.

The 15-week course met for one hour and fifteen minutes each Tuesday and Thursday. Suppose that Chapter 9 from the text had been lectured on the prior Thursday. On the following Tuesday then:

Table 1.   Factors Used in Determining Grades and Points Assigned to Each Factor

Factor	Percentage	Points
Individual homework effort	10	100
Individual mini-test performance	20	200
Group mini-test performance	20	200
Group comprehensive problem performance	10	100
Individual exam performance	10 each	300
Peer evaluation of individual's effort	10	100
Total	100	1,000

1. At the beginning of class, students submitted individual homework on Chapter 9 and a group solution to a comprehensive problem on Chapter 8 (5 minutes).
2. Students completed an inclass, individual mini-test on Chapter 9 (40 minutes).
3. Students redid inclass, as a group, the mini-test on Chapter 9 (45 minutes).

On the second class meeting of the week (Thursday), the following time line of activities would have been followed:

1. Individual and group mini-tests on Chapter 9 and comprehensive group problem solutions from Chapter 8 were returned to students. Instructor discussed difficult mini-test questions (10 minutes).
2. Students worked in their group to prepare a solution to a comprehensive problem from Chapter 9 (35 minutes).
3. Instructor lectured on Chapter 10 (30 minutes).

Of note in the above sequence of course activities followed each week is the fact that approximately half of the total, in-class time was devoted to cooperative-learning activities.

Measuring Individual Achievement

The first individual achievement measure of students in these sections was their average score on each of twelve individual mini-tests. The average of the twelve individual mini-test scores operationally defined the achievement level of each student for every unit of material, *before any of the scheduled cooperative-learning activities on the unit of material covered by the individual mini-test took place.*

The second measure of individual achievement was the mean score on the three formal examinations. This measure reflected the achievement of students *after* they engaged in cooperative-learning activities, such as taking the group mini-tests and working as a group on the comprehensive problems.

Group and Individual Bonuses

Students in both sections were afforded the opportunity to earn up to 50 bonus points on each exam, which meant that over the three exams, students could earn up to a total of 150 bonus points or 15% toward their final grade. In the 46-student section, bonus points could be earned by all members of a group, if the group's average exam score was greater than the group's average individual mini-test scores over that segment of the material. Specifically, the information sheet given to the 46 students in this section contained the following description of how bonus points could be earned:

You and everyone in your group can earn bonus points to raise your grade. Here is how it works. For each percentage improvement in the group's individual average exam score over the group's average individual mini-test score, your group will earn bonus points. For example, suppose an exam covers five chapters and your group's average on that exam is 70%. Suppose further that your group's average individual mini-test score on the five individual mini-tests was 60%. Everyone in your group would earn 50 bonus points (i.e., (70-60) x 5 points) toward increasing their final grade.

In the section with 54 students, bonus points could be earned by a student if his or her exam score was greater than the student's average individual mini-test scores over that segment of the material. Specifically, the information sheet given to these students contained the following description of how bonus points could be earned:

You can earn bonus points to raise your grade. Here is how it works. For each percentage improvement in your exam score over your average individual mini-test score, you will earn bonus points. For example, suppose an exam covers five chapters and you score 70% on that exam. Suppose further that your average individual mini-test score on the five individual mini-tests was 60%. You would earn 50 bonus points (i.e., (70-60) x 5 points) toward increasing your final grade.

Thus the *only* difference between the two sections was the basis for earning bonus points, (i.e., group versus individual performance improvement).

The Effect on Achievement

To measure the relative effect on achievement from cooperative learning associated with group rewards (compared to only individual rewards), the *difference between the exam score and the average individual mini-test score* (i.e., DIFFERENCE = EXAM SCORE–AVERAGE INDIVIDUAL MINI-TEST SCORE) was calculated for each student in both sections. By adopting this relative measure, not only is the design simplified by eliminating the need for covariates of achievement, but possible control problems related to unintentional and unmeasured differences between the sections are also mitigated.

An example serves to illustrate the benefits of employing a relative measure of achievement. Suppose, for example, that in one section students happened to have higher GPAs or that the instructor was unintentionally more effective in his or her lectures. All things being equal, a difference in the criterion variable may be obtained and attributed to learning activities, incentives, or other variables if the criterion variable is defined as the cumulative exam performance difference between these two sections. However, if the criterion variable of a student's exam performance is first scaled by a measure of that student's achievement level following the lecture (which is presumed higher in the section in which the students had higher GPAs or where the instructor delivered the more effective lectures), the difference after other learning activities occurred should disappear in the exam performance (which is also presumed to be higher in one section). That is, instead of examining the exam performance of two sections based on comparing their means, this study looks at a student's achievement after cooperative-learning activities *relative* to or scaled by that student's achievement before the occurrence of cooperative-learning activities.

Hypothesis

The null hypothesis of this study is that there will be no statistically significant difference in the exam-to-mini-test differences between the two sections. That is, students in the section with group bonuses for performance on exams and the students in the section with individual bonuses for performance on exams are hypothesized to benefit equally (i.e., have the same exam-to-mini-test differences) from the group-learning experiences. The *alternate* hypothesis is that the students in the section with the group bonuses will benefit more from a cooperative-learning process, as measured by an exam-to-mini-test difference, than will their counterparts in the section with individual bonuses. In other words, the mean exam-to-mini-test difference of the students in the section with group bonuses will be statistically greater than the mean exam-to-mini-t st difference of the students in the section with individual bonuses.

RESULTS

To help explain the research context better, Table 2 and Table 3 present summary statistics for the group-related and individual evaluation dimensions, respectively.

The average individual mini-test score for students in the section with group and individual bonuses was 53.5% and 53.2%, respectively. As indicated in Table 4, these average individual mini-test scores are not statistically different for the two sections (i.e., $t = 0.13$; $p < 0.90$, $df = 98$), and is as expected when students are randomly assigned to sections in the course registration process.

The average individual exam score[1] for the students in the section with group and individual bonuses was 58.9% and 54.6%,[2] respectively, indicating that students in both sections scored higher on the exams than the individual mini-tests. Further, as indicated in Table 5, the average exam scores for the two sections are not statistically different at conventional levels and the observed sample size.

Finally, Table 6 presents results that indicate the *relative* difference in achievement for students in the two sections. The average exam-to-mini-test difference for the students in the section with group and individual bonuses was 5.39 and 1.41, respectively. If no *differential* achievement-level benefits

Table 2. Average Group Mini-test Scores and Group Comprehensive Problemperformance of Students in Sections with Group and with Individual Rewards

Section	Number of Groups	Mini-test Scores (Percentages) Mean (Standard Deviation)	Group Problem Scores (Percentages) Mean (Standard Deviation)
Group Rewards	9	75.32 (5.65)	90.33 (6.24)
Individual Rewards	11	77.01 (6.78)	87.20 (4.82)

Table 3. Average Homework and Peer Evaluation of Students in Sections with Group and with Individual Rewards

Section	Number of Students	Homework Scores (Percentages) Mean (Standard Deviation)	Peer Evaluation Scores (Percentages) Mean (Standard Deviation)
Group Rewards	46	83.14 (8.63)	85.16 (7.20)
Individual Rewards	54	82.67 (9.34)	81.75 (6.95)

Table 4. Average Individual Mini-test Scores of Students in Sections with Group and with Individual Rewards

Section	Number of Students	Mini-test Scores (Percentages) Mean	Standard Deviation	Standard Error of the Mean
Group Rewards	46	53.51[*]	12.91	1.729
Individual Rewards	54	53.17[*]	12.71	1.903
Total	100	53.33	12.74	

Note: [*]Difference between means: $t = 0.13$; $p < 0.90$, $df = 98$

had been obtained (i.e., no evidence to reject the null hypothesis), similarity in the exam-to-mini-test differences for students in the two sections would be the expected finding. Applying a t-test[3] to these ratios, however, causes rejection of the null hypothesis (i.e., that there will be no statistically significant difference in these exam-to-mini-test differences for the two sections) at approxi-

Table 5. Average Individual Exam Scores of Students in
Sections with Group and with Individual Rewards

| | | Exam Scores (Percentages) | | |
Section	Number of Students	Mean	Standard Deviation	Standard Error of the Mean
Group Rewards	46	58.91[*]	13.37	1.972
Individual Rewards	54	54.59[*]	12.99	1.768
Total	100	56.57	13.28	

Note: [*]Difference between means: $t = 1.64$; $p < 0.11$, $df = 98$

Table 6. Exam-to-mini-test Differences of Students in
Sections with Group and with Individual Rewards

| | | Exam-to-mini-test Differences | |
Section	Number of Students	Mean	Standard Deviation
Group Rewards	46	5.39[*]	8.20
Individual Rewards	54	1.41[*]	7.46
Total	100	3.24	7.84

Note: [*]Difference between means: $t = 2.53$; $p < 0.02$, $df = 98$

Table 7. Results of Estimating Mini-test Performance and
Group Rewards against Exam Performance

Variable	Coefficient	Std. Error	t
Constant	9.85	3.36	2.93[*]
Mini-test Score	.84	0.60	14.01[**]
Group Bonus	4.03	1.53	2.64[*]

Notes: Adjusted R^2 0.67
Standard Error   7.71
F Statistic   102.08
Degrees of freedom   2,97
[*]Statistically significance at $p < 0.01$.
[**]Statistically significance at $p < 0.001$.

mately the 0.05 level (i.e., $t = 2.53$; $p < 0.02$, $df = 98$).[4] Therefore, the conclusion can be drawn with approximately a 95% level of confidence that the students in the section with group bonuses achieved more after participation in cooperative-learning activities than their counterparts in the section with only individual bonuses.

In an attempt to estimate the size of the group reward effect on exam achievement, the variables of individual mini-test performance and bonus determination (i.e., individual = 0; group = 1) were regressed against the criterion variable of exam performance. Table 7 reveals that both variables were statistically significant (individual mini-test performance, $t = 14.01$; $p < 0.001$, and group rewards, $t = 2.64$; $p < 0.01$). Table 7 also shows that after accounting for differences in mini-test performance, students who were in the section with group bonuses achieved approximately 4% higher scores exams than students who were in the section with individual bonuses.

CONCLUSION

The results of this study suggest that achievement of introductory managerial accounting students can be favorably impacted if group bonuses are used in conjunction with cooperative-learning techniques. Although it is recognized that this study is limited by its small sample size, and the fact that a single university served as the setting for the experiment, the results appear to be unambiguous and consistent with those found by Ravenscroft et al. (1995) and Ciccotello et al. (1997). Additionally, the fact that this study's results contrast with Ravenscroft et al.'s (1997) may be attributable, in part, to the manner in which the group bonuses were operationalized. By providing group bonuses based on relative improvement in achievement of a student, students of all levels of ability may have now become energized to engage in cooperative-learning activities. This may have meant that the more able students, by taking responsibility for ensuring that the less able students mastered the material, put forth added effort to teach the less able students in their group. Correspondingly, because of the increased attention now given to the less able students by their more able group members, possibly the less able students felt obligated "not to let the group down" by working harder to master the material.

For instructors who have not incorporated rewards that reflect the performance of all members of the group into the cooperative-learning fabric of their classes, much of the potential impact on student achievement may have been lost. When, for example, group bonuses for improvement in performance are absent, the group may, in cooperative-learning activities, trade off "learning-by-all" for efficiency. With only individual bonuses for improvement in performance, task responsibilities are apt to be divided among the group's members, with the more able students focusing on the definition and the analysis of the problem while the less able students are relegated to carrying out the clerical activities associated with preparing the problem's solution. As a result, there is little, if any, explicit incentive for the more able students to transfer their knowledge about the problem solution to the less able students. Moreover, the less able students do not feel obligated to expend additional effort to understand how the problem was solved because, as they rationalize, their share of the work has been done, demonstrating to their peers that they aren't "free riders."

ACKNOWLEDGMENT

I wish to thank the current co-editor of Advances in Accounting Education, Edward Ketz and the former co-editor of Accounting Education: A Journal of Theory, Practice and Research, David Stout for their many insightful comments and suggestions. My research also benefited greatly from the many helpful comments of four anonymous reviewers.

NOTES

1. These exam scores do not include any bonus points awarded.

2. Although average exam scores seem low, it should be noted that the GPA in this course traditionally is around 2.0. Moreover, because of the other grade components (i.e., group mini-tests, homework, group problems, and peer evaluations), students were able to earn an average final grade of 2.1 in these sections without having to grade on a curve.

3. A one-tailed t-test is appropriate here because the alternative hypothesis is directional.

4. It is legitimate to question the degree to which the distribution of the exam-to-mini-test differences in the two sections deviates from a normal distribution, thereby affecting the confidence that might be placed in the statistical results. Accordingly, the cumulative proportions of the exam-to-mini-test differences of the two sections were plotted against the cumulative proportions of normal distribution. This graphical depiction indicated that the exam-to-mini-test differences of both sections were approximately normally distributed.

REFERENCES

Ciccotello, C., R. D'Amico, and C. Grant. 1997. An empirical examination of cooperative learning and student performance in managerial accounting. *Accounting Education: A Journal of Theory, Practice and Research* 2: 1-7.

Johnson, D., and R. Johnson. 1990. Cooperative learning and achievement. In *Cooperative Learning*, edited by S. Sharan. New York: Praeger Press.

Johnson, D., and R. Johnson. 1994. *Learning Together and Alone*, 4th ed. Needham Heights, MA: Allyn and Bacon.

Lindsay, R., and A. Ehrenburg. 1993. The design of replicated studies. *The American Statistician* 47: 217-228

Michaelsen, L., C. Jones, and W. Watson. 1993. Beyond groups and cooperation: Building high performance learning teams. *To Improve the Academy* 11: 127-145.

Ravenscroft, S., F. Buckless, G. McCombs, and G. Zuckerman. 1995. Incentives in student team learning: An experiment in cooperative group learning. *Issues in Accounting Education* 10: 97-109.

Ravenscroft, S., F. Buckless, and G. Zuckerman. 1997. Student team learning—Replication and extension. *Accounting Education: A Journal of Theory, Practice and Research* 2: 151-172.

Slavin, R. 1990. An introduction to cooperative learning research. In *Learning to Cooperate, Cooperating to Learn*, edited by R. Slavin et al. New York: Plenum Press.

Slavin, R. 1995. *Cooperative Learning: Theory, Research, and Practice*, 2nd ed. Needham Heights, MA: Allyn and Bacon.

Stout, D., and J. Rebele. 1996. Establishing a research agenda for accounting education. *Accounting Education: A Journal of Theory, Practice and Research* 1: 1-18.

EFFECTIVE TEACHING TECHNIQUES:
PERCEPTIONS OF ACCOUNTING FACULTY

James D. Stice and Kevin D. Stocks

ABSTRACT

Improving the quality of accounting education has been the focus of many discussions over the last five years. Much of this debate has centered on content and pedagogical issues. Changes in accounting education have focused on various issues including the sequencing and/or integrating of topics, incorporating group learning techniques, and using computer technology in the classroom. However, one potential improvement that has received little attention is improving the quality of the teaching. To address this issue, we survey faculty with an open-ended questionnaire identifying factors affecting the perceived effectiveness of a teacher. Eighty-seven faculty identify 52 factors influencing perceived teacher effectiveness. A second group of 458 accounting professors then rate the 52 factors. Results identify important factors for creating a positive learning environment in five major categories: course content, classroom mechanics, teaching techniques, student involvement, and learning atmosphere. We also asked respondents to self-assess their teaching effectiveness. Based on this self-assessment, a subset of the 52 factors is identified that differentiates more effective teachers from less effective teachers.

Advances in Accounting Education, Volume 2, pages 179-191.

BECOMING A BETTER ACCOUNTING TEACHER

Changes in accounting curriculum have become commonplace. From introductory to advanced levels, accounting educators are reviewing and revising course content in an effort to improve the relevance of what is being taught. In addition, new learning methods are being incorporated into the classroom to assist students in becoming better prepared to face the challenges of a changing professional environment. Recent attention has turned to improving the quality of the teacher (AECC 1993; Rebele, Stout, and Hassell 1991; Stevens and Stevens 1992) reasoning that a better teacher will improve the learning experience independent of changes made to curriculum or pedagogy.

In this study, we examine factors affecting teaching effectiveness. Our purpose is to identify teaching techniques or factors that may improve the learning environment. Often, educators are encouraged to become better teachers but are given little guidance as to what they can do to improve. The results of this study provide educators with a menu of items that they can consider to improve their teaching effectiveness.

A large group of accounting faculty rated the effectiveness of 52 factors based on their own teaching experience. Additionally, we asked respondents to provide a self-assessment of their teaching. Based on this information, we partitioned responses and the results indicate that better teachers (as indicated by their self-assessment) incorporate and emphasize different teaching techniques than do their less effective counterparts. This information provides direction to all faculty who are striving to become more effective in the classroom.

The first section of this paper details the method used in collecting the data. The second section provides an analysis of the data to assess the perceived importance of the teaching techniques identified through the survey process. The third section partitions the respondents based on the self-assessment of their own teaching ability and provides an analysis of the relative emphasis placed on the teaching techniques. The last section offers conclusions and ideas for future research.

METHOD

Based on a review of the literature, we identified five general areas as being related to the teaching process (Bridges et al. 1971; Cahn 1982; Goldsmid et al. 1977; Kelley et al. 1991; Rebele et al. 1991). These areas include: (1) classroom management, (2) faculty/student relationships, (3) course content, (4) presentation skills, and (5) outside class activities. Using these five general categories, we constructed an open-ended questionnaire[1] to elicit feedback from a group of faculty regarding the specific factors within each category that improve their teaching process. An open-ended format was employed so as not to constrain respondents as they identified those factors that they found most effective. Respondents

Table 1. Factors Perceived by Respondents as Being Related to Teaching Effectiveness (87 respondents identified these factors)

COURSE CONTENT
  Discuss current events /literature
  Use real world examples and explanations
  Use the textbook mainly as a resource
  Talk with professionals to identify current topics
  Focus on textbook content
  Use outside speakers or guests
  Include material relevant to the CPA exam

CLASSROOM MECHANICS
  Organization:
    Clearly define student responsibilities
    Give course outlines to students on the first day of class
    Make as few changes to syllabus and grading as possible once
      class has started
    Use a detailed course outline
    Have an accounting lab
    Use seating charts
  Evaluation:
    Provide answer keys to exams
    Give quizzes
    Give exams in class, rather than in the testing center
  Running the class:
    Start class on time
    Identify objectives daily to students
    At end of each class period, provide a summary

LEARNING ATMOSPHERE
  In the class:
    Be enthusiastic
    Respond respectfully to student questions and viewpoints
    Have a sense of humor
    Project a real concern for students
    Learn student names
    Smile
    Create an informal classroom setting
  Outside of class:
    Encourage students to call or come to your office
    Interact with students during breaks
    Hold office hours regularly
    Be available in the room after class
    Be to class early
    Give students your home phone number
    Attend student activities, e.g., Beta Alpha Psi
    Keep notes at home to be able to respond to phone calls
    Use electronic office hours - E-mail
    Hold socials at the professor's home during semester

(continued)

Table 1   (Continued)

STUDENT INVOLVEMENT
   Ask for and accept student feedback
   Encourage student participation
   Use follow-up questioning
   Have students summarize relevant articles
   Have students help teach some topics
   Hold student interviews
   Encourage use of and volunteering for VITA or other service opportunities

TEACHING TECHNIQUES
   Move around the classroom
   Watch and learn from other good teachers
   Incorporate a variety of teaching techniques in the classroom
   Use the chalkboard
   Prepare overheads ahead of time
   Use cases
   Assign group projects
   Incorporate small group discussions
   Use multimedia technology

were asked, for each general category, to provide "a short description of those steps you have taken that you have found particularly effective" in improving the teaching process.

The participants in this study are obtained from the membership of the Teaching and Curriculum (T&C) Section of the American Accounting Association. The T&C Section has as its focus the promotion of teaching issues. T&C members are from all institutional types and from all faculty ranks.

We obtained a random selection of 1,000 members of the T&C Section. From this list, 100 accounting faculty were randomly selected and asked (via mail) to complete the open-ended questionnaire. Eighty-seven responses were received. Both authors reviewed each questionnaire and identified 52 different teaching techniques related to effective teaching. These 52 factors, partitioned by general category, are listed in Table 1. Because the purpose of this open-ended questionnaire was simply to identify teaching effectiveness factors, no attempt was made to track the number of responses for each factor or to try to gauge the relative importance of the various factors.

Having identified these 52 factors, we constructed a second questionnaire asking respondents to provide their perception as to the relative effectiveness of each of these factors. The remaining 900 faculty from the T&C Section mailing list received this questionnaire. Respondents were asked to identify a particular class upon which to base their responses. The authors made no effort to direct faculty to consider any specific course. The objective of this study is to identify factors impacting effective teaching in general, not in specific classes. The 52 activities were presented in random order with space provided for listing additional activities.

We then asked respondents to evaluate each activity in terms of "how effective or useful you have found it in creating a positive learning environment in teaching the class you identified..." using the scale: not effective; marginally effective; moderately effective; highly effective; and have not used. In addition, we asked respondents to provide a self assessment of the quality of their teaching using the scale: exceptional, excellent, very good, good, fair, or poor. We chose this scale because the terms represent those that often appear on teacher evaluation forms that students complete and that faculty review. Thus, we think that faculty are familiar with an evaluation scale of this type.

We left it to the faculty to define the levels of effectiveness as well as their own levels of teaching quality. Our effort in this study is to solicit data from faculty on what they feel makes their teaching effective. Hence we do not define survey measurement scales.

ANALYSIS

Of the 900 second-round questionnaires mailed, 458 were returned, yielding a response rate of almost 51%. Respondents identified the course type (i.e., financial, tax, etc.) and course level (i.e., introductory, intermediate, etc.) that they focused on in completing the questionnaire. An analysis of responses across course type and across course level (where sample size was large enough to allow comparison) resulted in no systematic differences.[2] As a result, in the analysis that follows, we group responses across course topic and course level. The fact that significant differences were not obtained across course topic and course level is of interest, because it indicates that factors that are perceived to improve teaching seem to be equally important across all topic areas and across all class levels.

In the following discussion, we examine each of the individual techniques, by category (see Table 1), and provide information as to the percentage of respondents that use a particular technique. We also investigate perceptions regarding the effectiveness of each technique in terms of creating a positive learning environment. To determine effectiveness, a mean score is computed for each factor using the following coding scheme: "Not Effective" is weighted 1; "Marginally Effective" is weighted a 2; "Moderately Effective" is weighted a 3; and "Highly Effective" is weighted a 4. Thus, the higher the mean score, the higher the level of effectiveness. If the respondent indicates that he or she has not employed a particular factor, it is not included in the computation of the mean score.

General Overview

Before reviewing specific results by category, some general observations are in order. First, we note a high correlation ($r = .84$) between a technique's perceived effectiveness and its use. We expected that those techniques perceived to be effective would tend to be applied more often. Second, rather than review results for each individual technique, we chose to focus on those techniques that are perceived to be most effective. Those techniques that are, on average, rated to be at least "moderately effective" (as indicated by a 3.0 rating) in creating a positive learning environment in the classroom are presented and discussed.

Course Content Category

Two course content factors received a mean score greater than 3.0: the use of real-world examples and explanations and discussing current events/literature (see Table 2, Panel A). Both of these factors address the importance of making the course content relevant to current business practice.

The use of outside speakers or guests is rated relatively low and utilized the least by the respondents. This result seems to indicate that faculty perceive the need for bringing current business implications of what they are presenting in the classroom, but not necessarily in the form of outside speakers. Another item of interest is that the inclusion of material relevant to the CPA exam (used by 81% of the respondents) is the lowest rated of the six factors.

Classroom Mechanics Category

The factors associated with classroom mechanics are summarized (Table 2, Panel B) into three areas: course organization, evaluation, and running the class. Four of the five factors affecting course organization have an average mean score above 3.0. From these factors, it appears that providing detailed information regarding the operation of a course and adhering to the information given are perceived by faculty as important factors relating to teacher effectiveness

With regard to evaluations, two of the three factors identified—giving exams in class rather than in a testing center and providing answer keys to exams—are rated above 3; however, both of these techniques are employed by only about 75% of respondents. When compared to the average "use rate" of 92% for those techniques considered "moderately effective," the application of these two techniques is not as common. However, viable explanations for the relatively low use of both are available. One obvious explanation relating to the "testing center" factor is that many colleges and universities do not have testing centers. With regards to providing answer keys to exams, many professors have a policy of keeping exams closed so as to be able to reuse exams in future periods.

Table 2. Mean Score and Relative Usage of Factors within Each Category

Overall Rank	Factor	Mean Score*	% of Respondents Who Have Used
	Panel A: Course Content		
9	Use real-world examples and explanations	3.54	94.8
23	Discuss current events/literature	3.29	97.8
34	Talk with professionals to identify current topics	2.97	81.7
36	Use the textbook mainly as a resource	2.92	80.8
38	Focus on textbook content	2.85	98.0
46	Use outside speakers or guests	2.72	59.4
47	Include material relevant to the CPA exam	2.71	81.4
	Panel B: Classroom Mechanics		
	Course Organization		
8	Clearly define student responsibilities	3.55	98.0
11	Give course outlines to students on the first day of class	3.50	94.1
15	Use a detailed course outline	3.44	93.2
22	Make as few changes to syllabus and grading as possible once the class has started	3.31	96.1
37	Have an accounting lab	2.87	44.5
	Evaluation		
31	Give exams in class rather than in the testing center	3.07	74.9
32	Provide answer keys to exams	3.06	71.6
33	Give quizzes	2.98	81.2
	Running the Class		
12	Start class on time	3.48	99.3
17	Identify objectives daily to students	3.37	85.8
35	At end of each class period, provide a summary	2.97	70.3
	Panel C: Teaching Techniques		
5	Incorporate a variety of teaching techniques in the classroom	3.56	94.3
14	Prepare overheads ahead of time	3.45	86.2
16	Watch and learn from other good teachers	3.39	83.0
19	Move around the classroom	3.35	94.5
24	Use the chalkboard	3.29	93.2
27	Incorporate small group discussions	3.19	75.3
28	Use cases	3.11	85.2
29	Assign group projects	3.08	78.0
41	Use multimedia technology	2.78	50.0
	Panel D: Student Involvement		
6	Encourage student participation	3.56	99.6
7	Ask for and accept student feedback	3.55	96.7
18	Use follow-up questioning	3.35	91.1
44	Have students summarize relevant articles	2.74	60.9
48	Encourage use of and volunteering for VITA	2.70	48.0
49	Have students help teach some topics	2.65	52.2
50	Hold student interviews	2.65	24.7

(continued)

Table 2   (Continued)

Overall Rank	Factor	Mean Score[*]	% of Respondents Who Have Used
	PANEL E: LEARNING ATMOSPHERE		
	In the class		
1	Be enthusiastic	3.81	97.8
2	Respond respectfully to student questions and viewpoints	3.80	99.8
3	Project a real concern for students	3.77	99.6
4	Learn student names	3.72	97.4
10	Have a sense of humor	3.53	99.1
13	Smile	3.47	96.5
20	Create an informal classroom setting	3.34	95.6
	Outside of Class		
21	Hold office hours regularly	3.32	97.4
25	Encourage students to call or come to your office	3.27	98.7
26	Interact with students during breaks	3.22	84.7
30	Be available in the room after class	3.07	94.3
39	Be to class early	2.84	83.6
42	Keep notes at home to be able to respond to phone calls	2.74	83.4
43	Attend student activities, e.g., Beta Alpha Psi	2.74	83.4
45	Give students your home phone number	2.73	59.2
51	Use electronic office hours - E-mail	2.55	33.2
52	Hold socials at the professor's home during semester	2.08	14.2

Notes: [*] Scale:
  1 = Not Effective
  2 = Marginally Effective
  3 = Moderately Effective
  4 = Highly Effective

Respondents rate as important efforts to start class on time (3.48) and to identify objectives daily to students (3.37). Again, it appears that attention to detailed communication is perceived to be important.

Teaching Techniques Category

Table 2, Panel C lists teaching factors identified in the study. All of the techniques listed except "use of multimedia technology" have a mean rating greater than 3.0. The factor receiving the highest rating, "incorporate a variety of teaching techniques in the classroom," is rated as the 5th highest factor in the study and also has one of the highest percentages of utilization. Note that three other factors—prepare overheads ahead of time, watch and learn from other good teachers, and move around the classroom—are rated in the top 20 factors identified in the study.

Student Involvement Category

Two factors in the student involvement category are rated very highly in this study (see Table 2, Panel D). Both "encourage student participation" and "ask for and accept student feedback" are factors focusing on two-way communication with students.

The other factors in this area have to do with specific activities that faculty undertake to increase student involvement. The relative rating of these factors correlates with their relative level of use.

Learning Atmosphere Category

The factors identified in the learning atmosphere category are separated into "in the class" and "outside of class" subcategories. The four highest rated factors associated with teaching effectiveness are in the "in the class" area (see Table 2, Panel E). It is interesting that all four of these factors relate to the student's perception of "how much the professor cares about me as an individual" as seen from a faculty's point of view.

All seven of the factors in the "in the class" area are rated among the top 20 identified in the study. The four factors in the outside of class area that rate above 3.0 all relate to faculty being available to students for one-on-one interaction. The ratings in the learning atmosphere category indicate their perceived importance in effective teaching.

IMPROVING TEACHER EFFECTIVENESS

To determine whether the perceived importance of individual activities is related to a teacher's self-reported level of effectiveness, the data are partitioned by the teachers' assessments of their teaching quality. The evaluations of the 52 factors by the excellent, very good, and good teachers are compared to the evaluations made by the exceptional teachers to identify significant differences. Comparisons are made only with the exceptional teachers.

The 52 techniques analyzed in this study are first ranked based on their overall mean score. A nonparametric Mann Whitney U-test is then performed for each factor. The responses from the excellent, very good, and good teachers are compared with those who self-assessed their teaching as exceptional. Of the 52 factors, there are only three statistically significant differences between those techniques ranked 21 through 52. As a result, those factors are not reported. Table 3 contains the results of these comparisons for activities ranked 1 through 20.

Table 3. A Comparison of Mean Scores of the Top 20 Factors
Across Self-Assessed Teaching Effectiveness Categories[**]

Rank	Factor	Exceptional	Excellent	Very Good	Good
1	Be enthusiastic	3.93	3.88	3.77[*]	3.49[*]
2	Respond respectfully to student questions and viewpoints	3.93	3.82	3.79[*]	3.57[*]
3	Have a sense of humor	3.89	3.57[*]	3.42[*]	3.38[*]
4	Project a real concern for students	3.84	3.85	3.72	3.47[*]
5	Learn student names	3.83	3.74	3.74	3.34[*]
6	Smile	3.77	3.52[*]	3.41[*]	3.03[*]
7	Move around the classroom	3.75	3.36[*]	3.27[*]	3.09[*]
8	Watch and learn from other good teachers	3.74	3.36[*]	3.35[*]	3.32[*]
9	Use real world examples and explanations	3.73	3.58	3.48[*]	3.42[*]
10	Ask for and accept student feedback	3.73	3.60	3.49[*]	3.30[*]
11	Encourage student participation	3.73	3.66	3.43[*]	3.34[*]
12	Clearly define student responsibilities	3.73	3.62	3.49[*]	3.14[*]
13	Incorporate a variety of teaching techniques in the classroom	3.70	3.63	3.47[*]	3.31[*]
14	Use follow-up questioning	3.65	3.40[*]	3.26[*]	3.06[*]
15	Start class on time	3.63	3.52	3.49	3.08[*]
16	Discuss current events/literature	3.61	3.27[*]	3.25[*]	3.11[*]
17	Give course outlines to students on the first day of class	3.61	3.52	3.49	3.16[*]
18	Create an informal classroom setting	3.60	3.33	3.36	3.03[*]
19	Encourage students to call of come to your office	3.49	3.38	3.18[*]	2.84[*]
20	Use the chalkboard	3.47	3.29	3.28	3.19

Notes: [*]Mann-Whitney U-test—significant difference at the .05 level.
[**]Each of the means from the three groups—excellent, very good, and good—was compared to the mean for the exceptional group. Only the top 20 factors are reported here as there were only three statistically significant differences among the remaining 32 factors.

Several interesting results are found in Table 3. When comparing exceptional teachers to good teachers, nineteen of the twenty factors are different. In every case, those differences result from the exceptional teacher rating the technique as more effective than did the good teacher. One might argue that the exceptional teachers simply rate everything as more effective than do the other groups of teachers. However, the fact that very few statistically significant differences are obtained with the remaining 32 (of 52) factors do not support this conclusion.

In every case where an activity is perceived differently by exceptional teachers versus excellent teachers, that activity is also perceived differently when compared to very good and good teachers. In no case is an activity significantly different for an excellent teacher and not significant for a very good or good teacher. These results suggest that self-reported exceptional teachers are emphasizing certain factors in the classroom to a higher degree than their colleagues who perceive themselves to be less effective teachers.

The results in Table 3 are based on respondents properly categorizing themselves according to their self-assessed effectiveness as teachers. Because the categories (i.e., exceptional, excellent, etc.) are not objective, respondents may misclassify themselves. If these misclassifications are random, the result would be to introduce noise into the statistical tests and bias against detecting significant differences. Also, the self assessment scale is presented as an interval scale. That is, the researchers assumed that the difference between an exceptional and an excellent teacher was the same as the difference between an excellent and a very good teacher, etc. If respondents do not make this assumption, it may be inappropriate to attach significance to the pattern of significance noted in the previous paragraph.[3]

It is possible that the above results are obtained because exceptional teachers simply rank the factors differently than do the other groups of teachers. That is, exceptional teachers may choose to implement a different set of factors than do excellent, very good, and good teachers. A nonparametric Kruskal-Wallis ranks test is performed comparing the ranking of the individual factors across the teaching categories. When comparing exceptional teachers with excellent and very good teachers, there is no significant difference ($p > .05$) between the ordering of the individual teaching techniques across the different groups. However, when comparing exceptional teachers with good teachers, a significant difference ($p < .01$) exists in the ordering of the factors.

Another possibility is that, given the number of statistical comparisons (3x20) being made in Table 3, we could erroneously conclude that some significant differences exist when, in fact, they do not. In other words, we should control for the possibility of a Type I error. To do this, each of the twenty factors in Table 3 is analyzed using a One-Way ANOVA controlling for the possibility of a type I error using Duncan's test (Kirk 1982). This procedure analyzes each factor across groups (perceived teaching effectiveness in this instance) and combines those groups where the means are not significantly different at a specified alpha level (probability of a type I error). For this test, the alpha level is set at .05. In only four instances do the results from this test differ from those displayed in Table 3.[4]

Overall, the results indicate that self-assessed exceptional, excellent, and very good teachers each perceive a similar group of techniques as able to influence teacher effectiveness. However, exceptional teachers emphasize certain techniques more strongly than do teachers in the other two groups.

CONCLUSION

Accounting educators have been challenged to change the way in which accounting is taught. We have identified numerous techniques that may be employed by
accounting faculty to improve their teaching effectiveness. Many of the most
important factors identified in this research relate to characteristics of individual
teachers. While what we teach may be important, this research indicates that how
we teach is important as well.

In comparing self-assessed exceptional teachers to other teachers, numerous
factors are identified that seem to be emphasized by exceptional teachers. This
result provides guidance to those wishing to improve their teaching. Many of the
items identified will not require a large amount of effort to be incorporated into a
teacher's portfolio of teaching techniques. For all those who have wanted to
become better teachers but have lacked direction in where to begin, the results of
this research provide accounting faculty with a starting point.

While this research identifies factors that may increase teacher effectiveness,
several questions remain unaddressed. Although we have identified factors
impacting perceived teaching effectiveness, we have not addressed any underlying attributes, such as personal compassion and the ability to communicate.
Another issue relates to class size. Are different teaching techniques more or less
effective in larger classes? This research investigates class level and class type but
leaves unanswered the question of class size. A third issue relates to the teacher's
self-assessment of teaching effectiveness. Obviously, an individual's self-assessment may be biased and an objective measure of teaching effectiveness would be
preferred. However, obtaining an objective measure of teaching effectiveness is
beyond the scope of this study. Finally, it would be interesting to compare
students' perceptions of factors relating to teacher effectiveness with those
assessments provide by faculty.

NOTES

1.   Readers who would like a copy of the research instruments employed in this study are encouraged to contact the authors.

2.   For example, Mann-Whitney U-tests comparing responses from introductory and intermediate
financial accounting resulted in only 3 of the 52 factors being significantly different across the two
groups at the .05 level. Comparing introductory and intermediate managerial accounting resulted in
only one significant difference. Comparing responses from tax and audit at the introductory level
resulted in only one significant difference at the 0.5 level.

3.   Statistical tests were conducted to determine if grouping the categories would influence the
results. Different groupings (e.g., combining excellent with exceptional and very good with good)
were made and the results indicate similar trends.

4.   In each of those four instances (factor numbers 1, 12, 13, and 20 from Table 3), the results from
the Duncan test indicated no significant difference between the exceptional, excellent, and very good
teachers. The results in Table 3 indicate that the responses from the very good teachers are
significantly different from those received from the exceptional teachers.

REFERENCES

Accounting Education Change Commission. 1993. Evaluating and rewarding effective teaching: Issues Statement No. 5. *Issues in Accounting Education* 8(Fall): 436-439.

Bridges, CM., W.B. Ware, B.B. Brown, and T. Greenwood. 1971. Characteristics of best and worst college teachers. *Science Education* 55: 545-553.

Cahn, S.M. 1982. The art of teaching: The essentials for classroom success. *American Educator* (Fall): 36-39.

Goldsmid, C.A., J.E. Gruber, and E.K. Wilson. 1977. Perceived attributes of superior teachers (PAST): An inquiry into the giving of teacher awards. *American Educational Research Journal* 14(Fall): 423-440.

Kelley C.A., J.S. Conant, and D.T. Smart. 1991. Master teaching revisited: Pursuing excellence from the students' perspective. *Journal of Marketing Education* (Summer): 1-10.

Kirk, R. E. 1982. *Experimental Design: Procedures for the Behavioral Sciences* (2nd ed., pp. 125-126). Brooks/Cole Publishing Company.

Rebele, I.E., D.E. Stout, and J.M. Hassell. 1991. A review of empirical research in accounting education: 1985-1991. *Journal of Accounting Education* 9: 167-231.

Stevens, K.T., and W.P. Stevens. 1992. Evidence on the extent of training in teaching and education research among accounting faculty. *Journal of Accounting Education,* pp. 125-141.

INTRODUCTION TO ACCOUNTING:
COMPETENCIES FOR NONACCOUNTING MAJORS

Mary S. Doucet, Thomas A. Doucet, and
Patricia A. Essex

ABSTRACT

This study presents the results of a survey of Finance, Management, and Marketing faculty concerning which topics from the first accounting courses are believed to be important to students in their respective disciplines. Respondents rated the extent of conceptual knowledge and technical ability as educational objectives that should be achieved for each of sixty topics. Analysis of the findings shows that respondents desire more conceptual knowledge of accounting topics than technical ability. Similarly, while a preference for user-oriented over preparer-oriented topics is clear, the preparer approach is still highly desired. The results show low priority placed on foundational topics (such as Accrual vs. Cash Accounting, Transaction Analysis, and Characteristics of Accounting Information) and strikingly little interest in integrative topics (such as those related to taxes, auditing, and systems). Finance faculty are found to differ from Management and Marketing faculty in their ratings. These results can be used for curriculum review but may be particularly beneficial as discussion points between accounting faculty and their nonaccounting faculty colleagues.

Advances in Accounting Education, Volume 2, pages 193-217.
Copyright © 2000 by JAI Press Inc.
All rights of reproduction in any form reserved.
ISBN: 0-7623-0515-0

INTRODUCTION

In its call for change in accounting education, the Accounting Education Change Commission (AECC) suggested that a greater focus should be placed on the needs of nonaccounting students in the first accounting course(s). The AECC (1992) elaborated by stating that "...the first course[s] in accounting should be an *introduction to accounting* rather than *introductory accounting*. It should be a rigorous course focusing on the relevance of accounting information to decision-making (use) as well as its source (preparation)" (emphasis in original, p. 250). As a result of the AECC initiative, many schools have changed or are considering changes to their first accounting course(s).

While these changes are well intentioned and likely will improve existing curricula, they are not always well supported by knowledge of the needs of business faculty who teach courses that follow the first classes in accounting. A number of factors might influence these needs, such as the discipline of the subsequent courses or the type of institution where the courses are taken. This study reports on a detailed survey of nonaccounting business faculty concerning the relative importance of accounting competencies for their courses and, as such, provides guidance for curriculum review. The survey covers a broad range of accounting material and, as often suggested for revamped curricula, includes topics from the financial, managerial, auditing, systems, and tax accounting areas.

The paper is structured as follows. In the next section the rationale behind this study is given, and then prior literature is reviewed. The third section describes the design of the study, followed by an analysis of the survey and a discussion of the results. The fifth section discusses limitations, and the last section discusses conclusions and provides suggestions for future research.

BACKGROUND

It is often presumed that a greater user orientation will serve the customers of the first accounting courses better than a preparer orientation. Typically, these customers are considered to be the students who take the course(s) and the employers who eventually will hire them. Another constituency that cannot be ignored, however, includes the faculty who teach in the other business disciplines such as finance, management, and marketing. These faculty rely on the first accounting course(s) to provide their students with certain competencies so that subsequent courses can build upon the foundation provided. The business faculty in these disciplines provide, in essence, the link between the topical coverage in the first accounting course(s) and the accounting knowledge, skills and abilities desired by employers in these other disciplines.

The first accounting courses can be viewed in terms of the model developed by Frederickson and Pratt (1995), which regards accounting education as a constrained optimization problem. In their model, the goal to be achieved is the minimization of the difference between the competencies demanded by those who employ accounting graduates and the competencies supplied by the educational process. In the present case, the model is applied to the first accounting course(s) rather than to the whole accounting program. Thus, the competencies demanded from customers who are internal to the organization (faculty in other business disciplines), rather than those who are external, are examined. This application of the model does not mean that future employers are not relevant nor that faculty in other disciplines are surrogates for future employers. Rather, other business faculty are treated here as internal customers whose ultimate goal is to serve the external user, the future employer.

An analysis of the accounting competencies demanded by faculty in other business disciplines should provide several benefits. First, a better understanding of the accounting knowledge and abilities needed by nonaccounting majors to succeed in their disciplines can be helpful to accounting faculty in structuring courses that better serve this constituency. More specifically, the information from this study will assist accounting faculty in determining what topics should be covered in the first accounting courses and whether the relative emphasis should be on achieving conceptual knowledge, developing technical ability, or both. As the AECC (1992) states, "The knowledge and skills provided by the first course[s] in accounting should facilitate subsequent learning even if the student takes no additional academic work in accounting or directly related disciplines" (p. 249).

Second, redesigning the first accounting course(s) with a focus on the needs of nonaccounting majors can be a positive step toward eliminating the reductionist view of the world that is often found in business schools. This view tends to simplify complex interdisciplinary problems, to the detriment of the student, by considering mainly the unique aspects of a particular discipline. To combat reductionism and to encourage redesigning our curricula to enhance the "scholarship of integration" Boyer (1992) suggests that:

> We need creative people who go beyond the isolated facts, who make connections across disciplines, who help shape a more coherent view of knowledge and a more integrated, more authentic view of life. And in our fragmented academic world, this task of integration becomes more urgent every day (p. 89).

Integration of accounting and nonaccounting material can be improved in both the accounting and the nonaccounting classrooms if all needs have been considered and communicated. In accounting courses, faculty might frame particular topics in terms of nonaccounting disciplines that will use those same topics at a later stage in the student's academic career. Also, nonaccounting business faculty, knowing that the basic accounting foundation has been provided, might design a more effective curriculum that reduces unnecessary redundancy and thereby provides greater educational value to students and employers.

Finally, designing the first accounting course(s) to better serve nonaccounting majors may help to attract more students who are considering accounting as their major field. The first accounting course is considered critical in attracting students to careers in accounting (AECC 1992), and a greater user focus should make the discipline more engaging and appealing. Thus, high-quality students increasingly may be attracted to the accounting profession. Furthermore, a greater user focus can provide a better understanding of the broad role of accounting in business, which should facilitate the later acquisition of procedural knowledge by accounting majors.

PRIOR LITERATURE

Several prior studies have examined issues related to the content of the first accounting courses. In fact, topical coverage in the first financial accounting course, as it pertains to the curriculum for accounting majors, was addressed well before the AECC initiative. Cherry and Reckers (1983) surveyed accounting faculty about the importance of topics in the first financial accounting course. In later studies, accounting faculty assessed what financial accounting topics should be covered in the first financial accounting course for both accounting majors and nonaccounting majors (Mintz and Cherry 1993), and nonaccounting business faculty rated the importance of 11 topical areas in the first financial accounting course for students majoring in their disciplines (Cherry and Mintz 1996).

This study differs in several significant respects from prior works. First, a more expansive list of topics and a more detailed description of these topics was provided. The expansive list of topics should assist accounting faculty in aligning topical coverage in the first accounting courses with the needs of faculty and students in other business disciplines. Additionally, the more detailed descriptions should minimize misunderstandings between accounting faculty and faculty in other business disciplines concerning topic coverage. This list of topics also should prove helpful in assessing the desired level of achievement of needed intellectual skills, an assessment that is more problematic when topics are aggregated.

In addition to more depth of description, the topic list provides a broader coverage of accounting topics. Managerial and other nonfinancial accounting topics are often essential for nonaccounting business majors. The only prior survey of nonaccounting faculty (Cherry and Mintz 1996) focused exclusively on financial accounting topics. The survey in this study also includes managerial, systems, tax, and auditing topics.

Respondents in this study rated the desired educational objective for each topic on two dimensions—conceptual knowledge and technical ability—to determine the depth of understanding that is expected. Conceptual knowledge is the ability to form an idea about a particular construct and ranges from minimal knowledge (such as simple recall or awareness) to extensive knowledge (such as that used in analysis, evaluation, and/or synthesis).[1] Technical ability refers to skill in applying existing knowledge to the solution of problems, and it also ranges from minimal to extensive for this survey. This rating scheme was used in a prior survey of accounting academics and professionals concerning topics from the accounting curriculum for accounting majors (Flaherty 1979).

DESIGN OF THE STUDY

Development of Survey Instrument

In developing the survey instrument, the questionnaire used in the American Accounting Association (AAA) monograph, *The Core of the Curriculum for Accounting Majors* (Flaherty 1979) served as a starting point. That study asked respondents to rate a number of topics for importance as well as for the educational objectives of conceptual knowledge and technical ability. It was presumed that if a topic was deemed to be unimportant by a respondent then achievement of any level of either conceptual knowledge or technical ability would be unimportant also. Further, evaluating each topic on three scales was reported to be "cognitively demanding" by faculty who reviewed the instrument. Therefore, the survey developed for the AAA was modified to have a "Not Important" check off box. Only if a topic was believed to be at least slightly important were respondents asked to rate desired levels of achieving the two educational objectives.

The topics included in the survey instrument were derived from several sources, starting with the master syllabi for the first accounting courses at our institution. Enhancements and simplifications were gleaned from other studies (Cherry and Mintz 1996; *The California Core Competency Model for the First Course in Accounting* July 1995; *University of Toledo Undergraduate Studies Committee Report on Core Curriculum* 1994; Mintz and Cherry 1993). After the topics were determined, they were sorted and organized into five broad categories: Analysis Tools, Reports and Reported Information, Systems and Processes, Common Topics, and Foundations.

Besides topical coverage, some studies have considered questions of a more general nature, such as whether communication and group skills should be addressed in the first accounting courses (e.g., Cherry and Mintz 1996; Mintz and Cherry 1993). Respondents have consistently rated such skills highly. Because the development of communication and group skills seems to be more related to pedagogical style (which is beyond the scope of this study) than to topical coverage, this study concentrates on identifying those intellectual skills necessary to prepare nonaccounting business majors for further study in their chosen discipline.

The survey instrument was pretested by faculty from other business disciplines at our institution. Based on their feedback, the instrument was revised to make it more user-friendly. Two noteworthy changes were necessary. The first was a simplification of the explanations of conceptual knowledge and technical ability. Longer definitions were deemed by these faculty to be unnecessary and more difficult to comprehend than the simpler ones provided here. The second revision was a rearrangement of the topical categories. The pretest faculty strongly opined that the response rate would suffer greatly unless faculty were asked to rate user topics (such as "Analysis Tools" or "Reports and Reported Information") before the more preparer-oriented background topics (such as "Foundations" or "Systems and Processes"). While this limits our ability to use factor analysis in analyzing the responses from this survey, these comments figured heavily in the decision not to have several versions of the instrument with questions rotated across the versions.

A second pretest was then conducted with only minor changes resulting. The final version of the instrument contained 60 topics representative of those covered in a two-semester introductory accounting sequence. Each topic was to be rated with respect to two educational objectives–conceptual knowledge and technical ability—that should be achieved in the first accounting courses by students who go on to major in the respondent's discipline. Five-point Likert scales were used with the endpoints of each scale being "minimal knowledge/ability" and "extensive knowledge/ability." Respondents who were not familiar with a topic, or who believed that a topic was not important, could indicate so with a check off box. Separate check off boxes were provided for "Not Familiar" and "Not Important" responses. A complete list of topics, along with means and standard deviations reported by discipline, is provided in the Appendix.

Survey Procedure

The survey was mailed to approximately 700 faculty members in each of three functional areas: Finance, Management, and Marketing. These faculty were selected randomly from Hasselback's directories for these disciplines.[2] The cover letter assured confidentiality of individual responses and encouraged participation. In addition to the topical items described above, participants were asked to answer 11 questions about themselves and their institutions. Several weeks after the initial response date, a follow-up survey questionnaire and cover letter were mailed to subjects who had not yet responded to the first request.

The overall response rate is 24%, but a number of surveys were too incomplete to be useable. Thus, the useable response rates are 21.3% for Finance faculty, 16.8% for Management faculty and 23.8% for Marketing faculty. The data were examined to determine whether a nonresponse bias exists by using the responses received from the follow-up mailing as surrogates for the nonrespondents and to test for differences between responses of this group and those of the first mailing (Oppenheim 1966, p. 34). Kruskal-Wallis tests of differences were performed on the 120 items for which nondemographic responses were requested. Five of the conceptual knowledge ratings from early respondents have differences that are

Table 1.   Demographic Information–Survey Respondents

	Finance	Management	Marketing	Total
Sample Size	710	708	701	2,119
Usable Responses (Useable Rate)	151 (21.3%)	119 (16.8%)	167 (23.8%)	437 (20.8%)
Respondents Holding:				
Masters degree Only	8 (5.3%)	5 (4.2%)	11 (6.6%)	24 (5.5%)
Doctorate degree	141 (93.4%)	110 (93.2%)	154 (92.2%)	405 (93%)
Average Years Teaching Experience	17.1	16.0	16.3	16.2
Respondents From:				
Four-Year Institutions	15 (9.9%)	14 (11.8%)	22 (13.3%)	51 (12.0%)
Masters-Level Institutions	98 (64.9%)	75 (63.0%)	111 (67.3%)	284 (65.0%)
Doctoral-Granting Institutions	38 (25.1%)	30 (25.2%)	32 (19.2%)	100 (22.8%)
Percentage Responding:				
Two accounting courses required	81.0	85.5	85.5	82.8
Course placed in sophomore year	82.8	68.1	77.9	74.0
Accounting courses are prerequisite to their courses	91.1	52.6	57.0	71.5
Satisfied with introduction to accounting at own institution	70.7	81.5	75.8	75.7
Business School Enrollments:				
Under 1,500	51.7	50.0	52.8	51.7
1,501 - 3,000	30.9	30.4	30.7	30.4
Over 3,000	18.1	19.6	16.6	17.9
University Enrollments				
Under 10,000	44.7	40.7	44.0	43.3
10,001 - 20,000	12.0	36.4	36.7	35.0
Over 20,000	23.3	22.8	19.3	21.7

statistically significant ($p = .05$) from those of late respondents, while only one rating of technical ability is significantly different. In a set of 120 items, on average, six items would be expected to differ between respondent groups simply by chance. Because the number of statistically significant differences found does not exceed those reasonably expected by chance alone, nonresponse bias is assumed to be minimal. A similar analysis was completed to test for nonresponse bias related to demographic characteristics, and no statistically significant differences were found.

Respondent demographics are presented in Table 1. The majority of the respondents hold doctorate degrees (more than 92% for each discipline), and the average teaching experience is in excess of 16 years. Other demographic information provided in Table 1 describes the respondents' institutions by degrees awarded (four-year, masters-level, or doctoral-granting) and by enrollments of the business school and university.

Nearly 83% of the respondents indicated that two first accounting courses are required of business students at their institutions, and approximately 74% stated that these courses are placed in the sophomore year. More than 91% of Finance faculty indicated that introductory accounting is prerequisite to the courses they teach, in contrast to only 52% of Management faculty and 57% of Marketing faculty. Since 85.5% of Management and Marketing faculty indicated that two accounting courses are required and 68.1% and 77.9% of Management and Marketing faculty, respectively, indicated that the placement of these courses is in the sophomore year, these courses are likely de facto prerequisites.

The level of dependence on these first courses could be related to faculty dissatisfaction with introductory accounting. Of the 379 faculty who responded to the question about whether they were satisfied with the first courses in accounting at their institutions, 29% of Finance faculty said they were not satisfied as compared to 18% of Management faculty and 24% of Marketing faculty. While this pattern indicates that the majority of nonaccounting business faculty are satisfied, it also indicates that there is room for improvement. Fine tuning of the first accounting courses in order to effect such improvements can be aided by knowledge of the competencies demanded. These competencies are explored in the next section.

RESULTS

The data were analyzed first to determine the competencies demanded by internal customers (i.e., Finance, Management and Marketing faculty); and, because these internal customers may have different demands, the analysis was discipline specific. Given that the broad range of institutions (four-year, masters, doctorate) included in our sample could imply diverse missions, demands made by internal customers might also differ by type of institution. Therefore, the data also were analyzed to determine whether type of institution influenced demanded competencies.

Thus, the first tests described here investigate the impact of DISCIPLINE, that is the functional discipline of the respondent, and INSTITUTION, defined by the highest degree awarded by the respondent's business school, on the mean ratings of two educational objectives: conceptual knowledge (CK) and technical ability (TA). Because respondents were asked to provide ratings of two educational objectives (OBJECTIVE) for each of 60 topics (TOPIC), a repeated measures analysis of variance (ANOVA) with repeated measures on one factor (educational objective) was performed. The other two factors included in the ANOVA were DISCIPLINE and INSTITUTION. The goal was to analyze the within-subjects interaction effects of DISCIPLINE and INSTITUTION on the two educational objectives (OBJECTIVE).

The ANOVA was first run treating all "Not Familiar" responses and all omitted responses as missing data.[3] Since the repeated-measures procedure drops observations where a response to any variable is missing, only 185 observations (out of 437) were allowed into the analysis. The results indicate that there are two statistically significant within-subject interactions–TOPIC by DISCIPLINE (F = 4.84, df = 18, $p < 0.01$) and TOPIC by OBJECTIVE (F = 5.21, df = 59, $p < 0.01$). No statistically significant within-subject interactions were found for OBJECTIVE by DISCIPLINE, OBJECTIVE by INSTITUTION, TOPIC by INSTITUTION, OBJECTIVE by TOPIC by DISCIPLINE, or OBJECTIVE by TOPIC by INSTITUTION. These results indicate that further analysis should concentrate on differences in responses based on discipline and educational objective. Further analysis of institutional effects is not warranted.[4]

TOPIC BY DISCIPLINE ANALYSIS

For each topic, respondents indicated the degrees of CK and TA important for success in their courses. "Not Important" responses were coded as zero (0) for both CK and TA. Thus, the scale ranged from 0 for "Not Important" responses to 5 for "Extensive Knowledge/Ability." The arithmetic means of the CK and the TA ratings for each topic were computed and arranged in sequential order to derive rankings of topics. This information is examined next.

Conceptual Knowledge (CK) Ratings

Every one of the 60 topics has a mean CK rating that is significantly different from 0 ($p < 0.01$) both overall and within each discipline.[5] Thus, conceptual knowledge of each topic is deemed at least somewhat important on an overall basis as well as by discipline.

The top 20 topics for each discipline are presented in Table 2. Some common preferences are evident. Related to financial accounting, the use of the three primary financial statements is considered very important. "Income Statement Use" and "Balance Sheet Use" are ranked in the top three for each discipline with "Statement of Cash Flows Use" near the tenth position. "Time Value of Money" is also ranked highly for all three disciplines (notwithstanding a few emphatic comments from Finance faculty that this topic belongs in their domain of teaching). Also notable are the topics of "Income Statement Preparation," "Statement of Cash Flows Preparation," and "Balance Sheet Preparation." While these topics have lower mean ratings than topics related to the *use* of financial statements, they are still considered to be very important by nonaccounting business faculty.

Although some topics are hard to categorize according to traditional financial/managerial classifications (such as "Return on Investment" discussed below), many faculty members find it convenient to speak of topics within this classification scheme. With regard to managerial accounting topics, all three disciplines consider conceptual knowledge of "Return on Investment" to be

Table 2. Top 20 Conceptual Knowledge (CK) Ratings by Discipline

Rank	Finance Topic (Survey Item #)	Mean	Management Topic (Survey Item #)	Mean	Marketing Topic (Survey Item #)	Mean
1	Income Statement Use (18)	4.30	Return on Investment (5)	3.86	Return on Investment (5)	4.17
2	Balance Sheet Use (16)	4.27	Income Statement Use (18)	3.82	Income Statement Use (18)	4.12
3	Time Value of Money (45)	4.26	Balance Sheet Use (16)	3.77	Balance Sheet Use (16)	3.96
4	Current Assets (37)	4.07	Trend Analysis (3)	3.65	Cost-Volume-Profit Analysis (11)	3.94
5	Corporate Equity (46)	4.03	Time Value of Money (45)	3.62	Time Value of Money (45)	3.69
6	Non-current Assets (38)	4.01	Nonfinancial Performance Measures (8)	3.59	Variance Analysis (12)	3.67
7	Long-term Liabilities (43)	4.00	Behavioral and Ethical Considerations (13)	3.59	Nonfinancial Performance Measures (8)	3.66
8	Depreciation and Amortization (39)	3.98	Cost-Volume-Profit Analysis (11)	3.56	Ratio Analysis (1)	3.63
9	Return on Investment (5)	3.92	Statement of Cash Flows Use (20)	3.52	Trend Analysis (3)	3.62
10	Statement of Cash Flows Use (20)	3.90	Ratio Analysis (1)	3.51	Current Assets (37)	3.57
11	Earnings per Share (4)	3.85	Variance Analysis (12)	3.37	Statement of Cash Flows Use (20)	3.52
12	Short-term Liabilities (42)	3.83	Current Assets (37)	3.37	Noncurrent Assets (37)	3.48
13	Income Statement Preparation (17)	3.83	Statement of Cash Flows Preparation (19)	3.34	Income Statement Preparation (17)	3.41
14	Statement of Cash Flows Preparation (19)	3.80	Income Statement Preparation (17)	3.30	Behavioral and Ethical Considerations (13)	3.39
15	Balance Sheet Preparation (15)	3.77	Earnings per Share (4)	3.29	Divisional Performance (6)	3.39
16	Ratio Analysis (1)	3.75	Noncurrent Assets (38)	3.28	Quality Costs (5)	3.34
17	Impact of Accounting Policy Choices (10)	3.75	Balance Sheet Preparation (15)	3.26	Depreciation and Amortization (39)	3.30
18	Capital Budgeting (34)	3.54	Accounting Information for Decisions (54)	3.26	Accounting Information for Decisions (54)	3.27
19	Forms of Business (59)	3.54	Capital Budgeting (34)	3.25	Statement of Cash Flows Preparation (19)	3.26
20	Accrual Vs. Cash Accounting (26)	3.41	Quality Costs (50)	3.23	Forms of Business (59)	3.23

Legend: Responses of "Not Important" were coded as (0) for computing the above means. Coding of other responses was based on a Likert scale where the endpoints were labeled as: (1) = Minimal Knowledge/Ability and (5) = Extensive Knowledge/Ability

very important. In fact, it is the top-ranked item for Management and Marketing faculty. However, the mean ratings of other managerial accounting topics vary across disciplines.

At a time when some accounting programs are making changes to the first accounting courses to include the other functional accounting areas, it is interesting to note that no auditing, systems, or tax accounting topics are ranked highly.

There appears to be some agreement among Finance, Management, and Marketing faculty about the particular accounting topics for which they believe their students should acquire higher levels of conceptual knowledge. Of the top 20 ranked CK ratings, there are ten items which all three disciplines indicate require these higher levels of conceptual knowledge. These ten items are "Income Statement Use," "Balance Sheet Use," "Time Value of Money," "Current Assets," "Noncurrent Assets," "Return on Investment," "Statement of Cash Flows Use," "Income Statement Preparation," "Statement of Cash Flows Preparation," and "Ratio Analysis." However, there are discipline-related differences among the rankings, particularly due to responses from Finance faculty. The Finance faculty's top 20 rankings do not include "Trend Analysis," "Nonfinancial Performance Measures," "Cost-Volume-Profit Analysis," "Variance Analysis," "Behavioral and Ethical Considerations," "Quality Costs," and "Accounting Information Requirements for Decisions." Most of these are traditional managerial topics and are ranked highly by both Management and Marketing faculty. "The Impact of Accounting Policy Choices," "Accrual vs. Cash Accounting," "Short-term Liabilities," "Long-term Liabilities," "Corporate Equity," and "Forms of Business," which relate to cash analysis and funding sources, are ranked more highly by Finance than Management and Marketing faculty. However, the results of a Friedman's Rank ANOVA test for consistency of ranks indicates that there are no statistically significant differences across disciplines in CK rankings.

Of particular interest may be those items at the bottom of the ranked list of 60 topics. Presented in Table 3 are the ten items that are ranked lowest by each discipline. Four items are ranked in the bottom ten for all three disciplines: "Design of Accounting Systems," "Types of Audit Opinions," "Internal Control Aspects of Accounting Systems," and "Economic and Social Purposes of Current Tax Structure." Again, the rankings for Finance faculty are unique for several topics. Within the lowest ten, only the rankings for Finance faculty include "Nonstatistical Variance Analysis," "Approaches to Accumulating Costs," "Activity-Based Costing," "Social Accounting," and "Quality Costs." Several items that are ranked lowest by Management and Marketing faculty are not ranked quite as low by Finance faculty; specifically, "Tax Returns and Schedules," "Treasury Stock," "Types of Taxes," "U.S. Reporting Standards and Regulatory Influences," and "Characteristics of Accounting Information." Management was the only discipline where "Residual Income" ranked in the lowest ten.

Table 3. Lowest Ten Conceptual Knowledge (CK) Ratings by Discipline

	Finance		Management		Marketing	
Rank	Topic (Survey Item #)	Mean	Topic (Survey Item #)	Mean	Topic (Survey Item #)	Mean
60	Design of Accounting Systems (35)	1.91	Types of Audit Opinions (52)	1.94	Design of Accounting Systems (35)	1.68
59	Internal Control Aspects (53)	2.01	Tax Returns and Schedules (23)	2.02	Types of Audit Opinions (52)	1.72
58	Quality Costs (50)	2.01	U.S. Reporting Standars and Regulatory Influences (55)	2.07	Tax Returns and Schedules (23)	1.85
57	Statistical Cost Estimation (31)	2.04	Economic and Social Purposes of Tax Structure (58)	2.18	U.S. Reporting Standards and Regulatory Influences (55)	1.91
56	Social Accounting (36)	2.05	Treasury Stock (47)	2.23	Economic and Social Purposes of Tax Structure (58)	1.94
55	U.S. Reporting Standards and Regulatory Influences (55)	2.11	Design of Accounting Systems (35)	2.25	Internal Control Aspects (53)	1.98
54	Types of Audit Opinions (52)	2.12	Types of Taxes (51)	2.30	Treasury Stock (47)	2.12
53	Activity-Based Costing (30)	2.13	Residual Income (7)	2.30	Characteristics of Accounting Information (56)	2.30
52	Nonstatistical Variance Analysis (9)	2.14	Characteristics of Accounting Information (56)	2.44	Statistical Cost Estimation (31)	2.37
51	Economic & Social Purposes of Tax Structure (58)	2.17	Internal Control Aspects (53)	2.47	Types of Taxes (51)	2.39

Legend: Responses of "Not Important" were coded as (0) for computing the above means. Coding of other responses was based on a Likert scale where the endpoints were labeled as: (1) = Minimal Knowledge/Ability and (5) = Extensive Knowledge/Ability

Technical Ability (TA) Ratings

As with the mean CK ratings, "Not Important" responses were coded as zero and included in the averages of TA ratings. All mean TA ratings are significantly different from 0 ($p < 0.01$).[6] Thus, TA for each topic is deemed at least somewhat important on an overall basis as well as for each discipline.

The top 20 TA topics reflect very closely the rankings of CK topics. Although the order is somewhat different, the same topics are ranked in the top 20 for TA as for CK with only a few exceptions that occur in the Marketing rankings. Similarly, the ten lowest-ranked topics for both TA and CK are the same across disciplines.

Upon closer inspection, unlike the findings for CK, the results of a Friedman's Rank ANOVA test for consistency of ranks indicates that there are statistically significant differences ($p < 0.01$) across disciplines in TA rankings. Tukey's Studentized Range (HSD) test for TA rankings indicates that Finance faculty rankings differ ($p < 0.05$) significantly from the rankings of the Management and Marketing faculty. It appears that overall Finance faculty rankings on TA are higher than Management and Marketing faculty rankings on TA.

Multiple-Comparison Analysis of Topic by Discipline

Although the rankings for TA are similar to those for CK, there are differences in the mean ratings for some topics. Tukey's Studentized Range (HSD) test of multiple comparisons was used to investigate whether these differences are statistically significant between disciplines. For each topic, the test was performed on mean CK ratings between disciplines, on a pair-by-pair basis, and then again for mean TA ratings between disciplines. There are statistically significant differences ($p < 0.05$) for the CK or TA mean ratings between at least two disciplines for 46 of the 60 topics respondents were asked to rate (see Table 4).[7]

The majority of these differences arise because the ratings of the Finance faculty differ from those of the Management faculty, the Marketing faculty, or both. For 19 topics, the mean ratings of the Finance faculty for CK and TA are significantly higher than the respective ratings of both the Management and Marketing faculties. With one notable exception (Capital Budgeting), all the topics for which this pattern holds true typically are considered to be financial accounting topics. Note that among the topics for which Finance faculty rate both CK and TA more highly are the preparation of the three primary financial statements. The Finance faculty's mean ratings are also significantly higher than those of the Management faculty for both educational objectives related to "Balance Sheet Use" and "Income Statement Use." Additionally, the Finance faculty's mean ratings are significantly higher than those of the Marketing faculty for both educational objectives related to "Statement of Cash Flows Use" and are significantly higher than Management faculty's mean TA rating of "Statement of Cash Flow Use."

Table 4. Statistically Significant Differences in Topic Ratings
Between Paired Disciplines

		Sample Size		Paired Disciplines		
	Topic	CK	TA	Finance and Management	Finance and Marketing	Management and Marketing
3	Trend Analysis	417	407	Both	CK	NS
4	Earnings-per-share	421	414	Both	Both	NS
5	Return on Investment	429	420	NS	NS	TA
6	Divisional Performance	394	387	NS	Both	NS
7	Residual Income	364	358	Both	Both	NS
8	Nonfinancial Performance Measures	399	388	Both	Both	NS
9	Nonstatistical Variance Analysis	370	364	NS	Both	Both
10	Impact of Accounting Policy Choices	410	404	Both	Both	NS
11	Cost-Volume-Profit Analysis	412	408	NS	Both	Both
12	Variance Analysis	413	411	CK	Both	NS
13	Behavioral and Ethical Considerations	422	409	TA	NS	NS
14	SEC Reports	409	404	Both	CK	NS
15	Balance Sheet Preparation	423	419	Both	Both	NS
16	Balance Sheet Use	426	419	Both	NS	NS
17	Income Statement Preparation	424	421	Both	Both	NS
18	Income Statement Use	426	419	Both	NS	TA
19	Statement of Cash Flows Preparation	425	419	Both	Both	NS
20	Statement of Cash Flows Use	415	409	TA	Both	NS
21	Flexible Budgets	361	351	NS	Both	NS
23	Tax Returns and Schedules	405	399	NS	CK	NS
26	Accrual versus Cash Accounting	393	385	Both	Both	NS
27	Approaches to Accumulating Costs	384	380	CK	CK	NS
28	Overhead Allocation and Control	403	395	CK	Both	NS
29	Variable Costing	392	385	CK	Both	NS
30	Activity-Based Costing	383	375	Both	Both	NS
31	Statistical Cost Estimation	374	371	CK	NS	NS
32	Responsibility Accounting	394	387	CK	Both	NS
33	Cash Management Practices	408	387	NS	Both	NS
34	Capital Budgeting	413	404	Both	Both	NS
35	Design of Accounting Systems	398	392	NS	NS	TA
36	Social Accounting	408	398	Both	CK	NS
37	Current Assets	420	413	Both	Both	NS
38	Noncurrent Assets	419	412	Both	Both	NS
39	Depreciation & Amortization	419	412	Both	Both	NS
42	Short-term Liabilities	416	406	Both	Both	NS
43	Long-term Liabilities	416	409	Both	Both	NS
44	Contingent Liabilities	395	387	Both	Both	NS
45	Time Value of Money	417	411	Both	Both	NS

(continued)

Table 4 (Continued)

	Topic	Sample Size		Paired Disciplines		
				Finance and Management	Finance and Marketing	Management and Marketing
		CK	TA			
46	Corporate Equity	417	411	Both	Both	NS
47	Treasury Stock	406	402	Both	Both	NS
48	Standard Costs	390	383	Both	CK	NS
49	Transfer Pricing	391	385	NS	Both	NS
50	Quality Costs	394	386	Both	Both	NS
51	Types of Taxes	409	392	TA	TA	NS
52	Types of Audit Opinions	399	392	NS	TA	NS
53	Internal Control Aspects	396	390	CK	NS	NS
55	U.S. Reporting Standards/Regulatory Influences	405	394	Both	Both	NS
56	Characteristics of Accounting Information	407	399	Both	Both	NS
58	Economic and Social Purposes of Tax Structure	406	393	NS	TA	NS
59	Forms of Business	419	404	TA	TA	NS
60	Business Cycles	414	399	NS	CK	NS

Legend:

Both = Statistically significant differences in both the mean CK and the mean TA ratings
CK = A statistically significant difference in the mean CK rating only
NS = No statistically significant difference in either the mean CK rating or the mean TA rating
TA = A statistically significant difference in the mean TA rating only
Only differences significant at least at the 0.05 level are reported. Significance was determined by both the Tukey's Studentized Range (HSD) Test and the Bonferroni (Dunn) t-test. See the appendix for actual mean ratings.

Most of the topics for which the Finance faculty's ratings are significantly lower than those of the Management or Marketing faculty are traditionally considered managerial topics. Two notable exceptions are "Trend Analysis" and "Social Accounting," which the Finance faculty rates significantly lower than the Management faculty for both educational objectives and significantly lower than Marketing faculty for the CK educational objective.

As shown in Table 4, statistically significant differences between mean ratings from the Management and Marketing faculties occur for only five topics. For two of these ("Nonstatistical Variance Analysis" and "Cost-Volume-Profit Analysis"), the Marketing faculty rated the topics higher than the Management and the Finance faculties for both educational objectives. The Marketing faculty also rated "Return on Investment" and "Income Statement Use" significantly higher than Management faculty for the TA educational objective. The only topic which the Management faculty rated higher than the Marketing faculty is the "Design of Accounting Systems" for the CK educational objective.

Multiple-Comparison Analysis with Topical Categories

Another multiple-comparison analysis was performed on the mean ratings of topics as classified in five categories: Analysis Tools, Reports and Reported Information, Systems and Processes, Common Topics and Foundations (see Appendix for category components). These categories were delineated on the survey questionnaire, and differences across disciplines might be attributable to these categories.

The pattern that emerges here is consistent with the findings from the topic-by-topic analysis discussed above. That is, the Finance faculty's mean ratings by category, for one or both educational objectives, are different from the respective mean ratings of the Management faculty, the Marketing faculty, or both. For the category of Reports and Reported Information, the Finance faculty's mean ratings are higher than those of the Management faculty for both educational objectives. For the categories of Common Topics and Foundations, the Finance faculty's mean TA ratings are higher than those of both the Management and Marketing faculties. There are no statistically significant differences across disciplines for the category of Systems and Processes, which consists of topics that were generally not highly ranked.

TOPIC BY OBJECTIVE ANALYSIS

While there are many differences in the mean ratings across disciplines, one striking result is that, for all topics, technical ability is rated lower than conceptual knowledge. That is, the mean CK rating minus the mean TA rating (hereafter called the CK premium) is always greater than 0. However, the overall ANOVA interaction effect of TOPIC by OBJECTIVE ($F = 5.21$, df = 59, $p < 0.01$) indicates that statistically significant differences in the magnitude of the CK premium occur across topics.

Based on a multiple-comparison analysis, the CK premium is greater for "softer" topics (such as "Impact of Accounting Policy Choices," "Behavioral and Ethical Considerations," "Social Accounting," "U.S. Reporting Standards and Regulatory Influences," "Characteristics of Accounting Information," and "Forms of Business") than for the "hard" computational topics (such as "Ratio Analysis," "Trend Analysis," "Earnings-per-Share," "Return on Investment," "Residual Income," "Nonstatistical Variance Analysis," and "Cost-Volume-Profit Analysis"). More interesting is that the CK premium is greater for "Accrual versus Cash Accounting," "Overhead Allocation and Control," "Responsibility Accounting," "Goodwill," and "Types of Taxes" than for those topics identified above as computational topics.[8]

DISCUSSION OF RESULTS

The results of this study provide a better understanding of the competencies demanded of the first accounting courses. One prior study (Cherry and Mintz 1996) asked nonaccounting business faculty to rate the importance of 11 topical areas in the first financial accounting course for students majoring in their disciplines. They found that Finance faculty's response patterns differed from those of other nonaccounting business faculty because Finance faculty gave highest priority to balance sheets and income statements and because they were more interested in having the statement of cash flows covered. To further explore the educational requirements of our nonaccounting business colleagues, in this study we greatly expanded the list of topical coverage, asked participants to rate two different educational objectives (conceptual knowledge and technical ability), and analyzed responses by discipline. Data provided by the nonaccounting faculty in this study can assist accounting faculty in gauging the degree to which these two educational objectives should be achieved by nonaccounting majors.

While a number of topics appear consistently in the top 20 rankings for Finance, Management, and Marketing faculties, a pattern of difference, similar to that of the Cherry and Mintz (1996) study, between the Finance faculty and the other two disciplines is evident. This pattern results from Finance faculty's providing generally higher ratings for financial accounting topics than for managerial accounting topics. This pattern of difference cannot be ignored and serves to highlight the need for greater communication with our colleagues in other disciplines.

Although overall the rankings for preparation and use of financial statements are high, topics that underlie these financial statement skills rank considerably lower. The respondents gave less importance to items such as "Accrual vs. Cash Accounting" (except for Finance faculty), "Transaction Analysis," "Double-Entry System," and "Characteristics of Accounting Information." Without coverage of these topics, successful acquisition of conceptual knowledge and technical abilities related to financial statements is likely to be difficult. Perhaps there is an assumption on the part of the nonaccounting faculty that whatever foundation is necessary to cover the higher-ranked topics successfully will be provided, or perhaps this finding is a further indication of the need for communication across disciplines.

The competencies demanded by nonaccounting faculty do not appear to include auditing, systems, or tax topics. Topics in these areas are not rated highly by the respondents to this survey. In fact, "Types of Audit Opinions" and "Internal Control Aspects of Accounting Systems" appear in the bottom ten ranked topics for all three disciplines. "Types of Taxes" and "Tax Returns and Schedules" appear in the bottom ten ranked topics for Management and Marketing faculty. While these rankings may be disturbing, especially to accounting faculty in these areas, we must recognize that some of these topics should still be covered in the first accounting courses because they benefit the general education of students.

Additionally, while the needs of nonaccounting business faculty have been the focus of this study, accounting faculty must not neglect their directly related constituencies: faculty who teach upper-level accounting courses and professionals.

LIMITATIONS

There are several limitations to this study. First, the survey was restricted to only three business disciplines, which seemed justified because these three disciplines account for the majority of nonaccounting business majors. Second, while the useable response rates for Finance and Marketing were greater than 20%, the useable response rate for Management was slightly less than 17%. Unfortunately, a number of Management faculty responded that they teach in graduate programs only and consequently they did not complete the survey. A third limitation relates to the constant ordering of the topics on all surveys. This was done in accordance with advice from the pretest faculty, but an order effect is possible. Fourth, as with any study where a specialized language (such as accounting) is used, some of the topic descriptions may not have been detailed enough for respondents to recognize. Finally, while the list of topics was expanded from prior works, it still may omit potentially important topics for the first accounting courses.

CONCLUSION

Just as the external auditor is required to consider the needs of all parties relying on the audit report, accounting faculty should consider the needs of all constituencies, including those customers who are internal to the organization and whose needs may vary. This consideration requires that accounting faculty communicate with faculty in other disciplines about the accounting needs of their students in terms of the breadth of topical coverage and the depth of the desired conceptual understanding and technical ability for these topics. While the results of this study provide a more comprehensive set of talking points for communication with our nonaccounting colleagues, accounting faculty have to consider and balance the sometimes competing needs of these internal constituents at their respective institutions. Communicating with our colleagues in other disciplines about what they need in the first accounting courses can help us provide a more integrated, authentic view of the business world in those courses and the ones that follow them.

Part of this more authentic view of the business world may include room for a less dichotomous stance on the user-versus-preparer debate. While respondents in this study rated financial statement use higher than financial statement preparation, acquisition of technical abilities for Balance Sheet, Income Statement, and Statement of Cash Flows preparation still ranked highly. Obviously, exclusion of preparation skills from the first courses may be ill advised. On the whole, the analysis indicates that nonaccounting business faculty might prefer a blending of the user and preparer approaches, with somewhat more emphasis on the *use* of accounting information.

Additional research related to the first accounting course(s) could produce useful results. The limitations in this study, as described above, provide opportunities for exploration. Other avenues for related future research include the investigation of questions such as the following: Are competencies demanded by nonaccounting faculty actually being supplied at their schools? Do schools with the biggest perceived deficiencies have communication patterns across faculties that differ from communication patterns at schools where the competencies demanded are better met? Do accounting faculty disseminate appropriate levels of information to nonaccounting colleagues about changes to the first courses, such as integration of other accounting functional areas and the rationale underlying such changes?

Topical coverage, educational objectives, and competencies provided by the first accounting courses must balance the needs of both internal and external constituencies. The purpose of this study was to take a more pronounced look at the needs of those constituents who are internal to the organization and whose voices may not have been well considered. The results provide accounting faculty with a more comprehensive opening dialog for communication with their nonaccounting colleagues as they seek to enhance the learning environment for all business students.

DATA AVAILABILITY

Data will be available for a nominal fee. Contact any author to initiate that process.

APPENDIX

	Conceptual Knowledge						Technical Ability					
	Finance		Management		Marketing		Finance		Management		Marketing	
	Mean	(Std.Dev)	Mean	(Std.Dev)	Mean	(Std.Dev)	Mean	(Std.Dev)	Mean	(Std.Dev)	Mean	(Std.Dev)
Financial Statement Analysis:												
1 Ratio analysis	3.75	(1.20)	3.51	(1.43)	3.63	(1.14)	3.53	(1.28)	3.31	(1.46)	3.38	(1.22)
2 Common-size analysis (e.g., vertical analysis)	2.91	(1.31)	2.76	(1.58)	2.70	(1.38)	2.73	(1.30)	2.56	(1.57)	2.48	(1.35)
3 Trend analysis	3.17	(1.29)	3.65	(1.27)	3.62	(1.13)	2.94	(1.27)	3.33	(1.34)	3.29	(1.25)
4 Earnings-per-share	3.85	(1.23)	3.29	(1.48)	3.19	(1.39)	3.61	(1.30)	2.98	(1.54)	2.89	(1.33)
Performance Evaluation Tools:												
5 Return on investment	3.92	(1.22)	3.86	(1.25)	4.17	(0.94)	3.67	(1.25)	3.42	(1.37)	3.89	(1.10)
6 Divisional performance	2.69	(1.43)	3.01	(1.52)	3.39	(1.34)	2.46	(1.38)	2.60	(1.49)	3.03	(1.37)
7 Residual income	3.09	(1.48)	2.30	(1.54)	2.65	(1.42)	2.80	(1.37)	2.04	(1.50)	2.34	(1.35)
8 Nonfinancial performance measures	2.42	(1.42)	3.59	(1.38)	3.66	(1.15)	2.12	(1.32)	3.18	(1.41)	3.29	(1.27)
9 Nonstatistical variance analysis	2.14	(1.29)	2.55	(1.56)	3.03	(1.30)	1.94	(1.23)	2.17	(1.41)	2.67	(1.31)
10 Impact of accounting policy choices (e.g., inventory valuation methods, depreciation methods)	3.75	(1.22)	3.09	(1.53)	3.22	(1.51)	3.16	(1.30)	2.50	(1.43)	2.66	(1.40)
11 Cost-volume-profit analysis	3.21	(1.38)	3.56	(1.22)	3.94	(1.19)	2.94	(1.39)	3.10	(1.39)	3.67	(1.31)
12 Variance analysis (e.g., budgeted vs. actual performance)	2.94	(1.32)	3.37	(1.56)	3.67	(1.27)	2.63	(1.29)	2.91	(1.55)	3.31	(1.36)
13 Behavioral and ethical considerations in the use of accounting information	3.17	(1.55)	3.58	(1.44)	3.39	(1.52)	2.54	(1.45)	3.12	(1.55)	2.82	(1.59)
Reports and reported information:												
14 SEC reports (i.e., 10-Ks, 10-Qs, proxy statements, etc.)	3.11	(1.35)	2.48	(1.63)	2.65	(1.63)	2.59	(1.38)	2.08	(1.52)	2.26	(1.52)
15 Balance sheet preparation	3.77	(1.36)	3.26	(1.54)	3.20	(1.46)	3.39	(1.46)	2.74	(1.49)	2.79	(1.50)
16 Balance sheet use	4.27	(1.01)	3.77	(1.28)	3.96	(1.12)	3.97	(1.18)	3.31	(1.46)	3.66	(1.25)
17 Income statement preparation	3.83	(1.31)	3.30	(1.56)	3.41	(1.38)	3.48	(1.41)	2.84	(1.55)	3.08	(1.43)

#	Topic											
18	Income statement use	4.30	(0.95)	3.82	(1.39)	4.12	(1.00)	4.00	(1.14)	3.41	(1.58)	3.81	(1.21)
19	Statement of cash flows preparation	3.80	(1.30)	3.34	(1.59)	3.26	(1.43)	3.51	(1.38)	2.88	(1.59)	2.87	(1.47)
20	Statement of cash flows use	3.90	(1.05)	3.52	(1.53)	3.52	(1.28)	3.63	(1.21)	3.14	(1.57)	3.22	(1.38)
21	Flexible budgets	2.23	(1.37)	2.67	(1.58)	2.75	(1.46)	1.92	(1.23)	2.20	(1.39)	2.36	(1.40)
22	Budgeted financial statements	2.62	(1.45)	2.71	(1.55)	2.89	(1.54)	2.36	(1.40)	2.28	(1.45)	2.59	(1.54)
23	Tax returns and accompanying schedules	2.27	(1.39)	2.02	(1.45)	1.85	(1.48)	1.85	(1.25)	1.59	(1.31)	1.55	(1.33)

Systems and Processes:

#	Topic											
24	Double-entry system	2.90	(1.71)	2.63	(1.74)	2.59	(1.74)	2.54	(1.67)	2.15	(1.63)	2.16	(1.56)
25	Transaction analysis	2.45	(1.54)	2.49	(1.62)	2.63	(1.63)	2.24	(1.49)	2.13	(1.55)	2.18	(1.47)
26	Accrual versus cash accounting	3.41	(1.40)	2.63	(1.68)	2.54	(1.71)	2.95	(1.41)	2.07	(1.47)	1.97	(1.49)
27	Approaches to accumulating costs (e.g., job order costing, process costing)	2.11	(1.46)	2.62	(1.47)	2.57	(1.66)	1.78	(1.29)	2.09	(1.33)	2.04	(1.47)
28	Overhead allocation and control	2.38	(1.65)	3.02	(1.43)	3.13	(1.49)	2.01	(1.48)	2.35	(1.30)	2.58	(1.42)
29	Variable costing	2.21	(1.56)	2.88	(1.48)	3.06	(1.56)	1.93	(1.44)	2.34	(1.38)	2.57	(1.50)
30	Activity-based costing	2.13	(1.57)	3.17	(1.48)	3.01	(1.67)	1.84	(1.41)	2.48	(1.39)	2.50	(1.56)
31	Statistical cost estimation	2.05	(1.47)	2.62	(1.42)	2.37	(1.57)	1.76	(1.35)	2.12	(1.34)	2.01	(1.41)
32	Responsibility accounting (e.g., cost, revenue, profit, and investment centers)	2.44	(1.60)	3.05	(1.48)	3.08	(1.60)	2.06	(1.47)	2.29	(1.32)	2.49	(1.48)
33	Cash management practices	3.21	(1.36)	2.96	(1.59)	2.79	(1.51)	2.89	(1.42)	2.46	(1.47)	2.38	(1.47)
34	Capital budgeting	3.75	(1.43)	3.25	(1.53)	2.99	(1.57)	3.55	(1.50)	2.78	(1.49)	2.55	(1.50)
35	Design of accounting systems	1.91	(1.43)	2.25	(1.57)	1.68	(1.39)	1.57	(1.22)	1.58	(1.28)	1.30	(1.12)
36	Social accounting (e.g., costs of meeting societal responsibilities, effects upon management decisions)	2.05	(1.50)	2.77	(1.60)	2.61	(1.58)	1.56	(1.22)	2.08	(1.34)	1.92	(1.37)

Common Topics:

#	Topic											
37	Current assets (e.g., cash, receivables, inventories, prepaid items)	4.07	(0.97)	3.37	(1.51)	3.57	(1.38)	3.77	(1.09)	2.92	(1.49)	3.01	(1.39)
38	Non-current assets (e.g., fixed assets, investments)	4.01	(0.96)	3.28	(1.53)	3.48	(1.41)	3.68	(1.11)	2.83	(1.50)	2.90	(1.40)
39	Depreciation, depletion, and amortization	3.98	(1.01)	3.04	(1.57)	3.30	(1.50)	3.65	(1.12)	2.58	(1.52)	2.72	(1.43)

(continued)

Appendix (Continued)

	Conceptual Knowledge						Technical Ability					
	Finance		Management		Marketing		Finance		Management		Marketing	
	Mean	(Std.Dev)	Mean	(Std.Dev)	Mean	(Std.Dev)	Mean	(Std.Dev)	Mean	(Std.Dev)	Mean	(Std.Dev)
40 Goodwill	3.25	(1.32)	2.89	(1.49)	3.21	(1.45)	2.71	(1.30)	2.32	(1.38)	2.59	(1.36)
41 Other intangible assets	2.81	(1.37)	2.60	(1.55)	2.83	(1.51)	2.39	(1.31)	2.17	(1.49)	2.23	(1.32)
42 Short-term liabilities	3.83	(1.00)	3.01	(1.52)	3.04	(1.50)	3.50	(1.12)	2.53	(1.49)	2.55	(1.40)
43 Long-term liabilities	4.00	(0.96)	3.05	(1.55)	3.09	(1.48)	3.66	(1.06)	2.58	(1.52)	2.56	(1.40)
44 Contingent liabilities	3.11	(1.30)	2.59	(1.63)	2.59	(1.56)	2.66	(1.27)	2.10	(1.44)	2.10	(1.38)
45 Time value of money	4.26	(1.18)	3.62	(1.53)	3.69	(1.51)	4.06	(1.33)	3.12	(1.61)	3.35	(1.50)
46 Corporate equity	4.03	(1.11)	3.05	(1.65)	3.02	(1.53)	3.72	(1.26)	2.60	(1.59)	2.54	(1.40)
47 Treasury stock	3.03	(1.31)	2.23	(1.58)	2.12	(1.46)	2.57	(1.31)	1.79	(1.38)	1.65	(1.22)
48 Standard costs	2.24	(1.41)	2.81	(1.56)	2.72	(1.53)	1.94	(1.35)	2.39	(1.50)	2.19	(1.39)
49 Transfer pricing	2.23	(1.43)	2.70	(1.54)	3.09	(1.57)	1.89	(1.26)	2.24	(1.44)	2.52	(1.49)
50 Quality costs (cost/benefit of quality products or services)	2.01	(1.37)	3.23	(1.53)	3.34	(1.46)	1.71	(1.18)	2.76	(1.48)	2.83	(1.38)
51 Types of taxes	2.69	(1.41)	2.30	(1.56)	2.39	(1.53)	2.23	(1.30)	1.75	(1.28)	1.72	(1.23)
52 Types of audit opinions	2.12	(1.50)	1.94	(1.62)	1.72	(1.44)	1.66	(1.25)	1.38	(1.22)	1.25	(1.13)
53 Internal control aspects of accounting systems	2.01	(1.53)	2.47	(1.72)	1.98	(1.49)	1.59	(1.28)	1.89	(1.41)	1.57	(1.29)
54 Accounting information requirements for decisions (such as pricing, lease vs. purchase, make vs. buy)	2.99	(1.52)	3.26	(1.63)	3.27	(1.55)	2.65	(1.48)	2.64	(1.59)	2.79	(1.56)
Foundations:												
55 U.S. reporting standards and regulatory influences (e.g., FASB, AICPA, SEC)	2.62	(1.46)	2.07	(1.56)	1.91	(1.50)	2.14	(1.35)	1.52	(1.29)	1.29	(1.11)
56 Characteristics of accounting information (e.g., historical cost, matching expenses to revenues, materiality, etc.)	2.96	(1.49)	2.44	(1.67)	2.30	(1.55)	2.41	(1.44)	1.83	(1.32)	1.65	(1.27)

#	Item												
57	Information economics (i.e., cost/benefit of information)	2.84	(1.41)	2.76	(1.65)	2.87	(1.56)	2.41	(1.40)	2.10	(1.47)	2.32	(1.49)
58	Economic and social purposes of current tax structure	2.17	(1.41)	2.18	(1.64)	1.94	(1.44)	1.85	(1.29)	1.63	(1.38)	1.42	(1.20)
59	Forms of business (corporations, partnerships, proprietorships)	3.54	(1.29)	3.22	(1.58)	3.23	(1.60)	3.06	(1.33)	2.53	(1.50)	2.60	(1.55)
60	Business cycles (acquisition, production, sales)	2.62	(1.35)	3.03	(1.60)	3.13	(1.55)	2.26	(1.32)	2.43	(1.47)	2.51	(1.51)

Legend: Responses of "Not Important" were coded as (0) for computing the above means. Coding of other responses was based on a Likert scale where the endpoints were labeled as: (1) = Minimal Knowledge/Ability and (5) = Extensive Knowledge/Ability

ACKNOWLEDGMENTS

This research was funded by a grant to the first author from the Ohio Society of CPAs and a grant to the second and third authors from the College of Business Administration at Bowling Green State University.

NOTES

1.   As used here, extensive knowledge is at the upper end of Bloom's taxonomy of educational objectives (Bloom 1956, 20-207). This terminology may be problematic to some since Bloom et al. describe knowledge as the lowest level of learning, and the term "knowledge" was used here in the description of the upper level of cognitive objectives. This practice was favored in order to provide a descriptor that did not require a great deal of explanation and to be consistent with Flaherty (1979).

2.   Professor Hasselback drew and supplied a random sample of at least 700 from each of the databases he maintains for the three functional disciplines.

3.   "Not Familiar" responses may have arisen because the respondent did not understand the meaning of the phrase used (but would have understood if it was stated differently) or because the respondent had never before encountered the topic. Consequently, a "Not Familiar" response could not be assumed to mean that a topic was not important. Unless specifically stated otherwise for a particular analysis, "Not Familiar" responses were treated as missing observations of the associated CK and TA ratings.

4.   A number of respondents did not supply ratings of both CK and TA for all 60 topics, and the statistical analysis reported in the text dropped such observations. (Over 90% of the respondents provided a CK rating for 50 of the 60 topics and a TA rating for 37 of the 60 topics. The only topic to receive less than a 75% response rate was the topic, "Common Size Statements," with which many respondents indicated that they were unfamiliar.) An alternative ANOVA was performed to verify that findings from the first analysis were not due to some anomalous characteristic of those repondents who provided both a CK and a TA rating for all 60 topics. For this second test, to prevent observations from being dropped, "Not Familiar" and omitted responses for CK and TA were assigned their respective median values. This approach allowed all responses to individual CK and TA ratings to be included even though one of the other variables from an observation might be missing.

The results of this second analysis support the findings of the first and suggest that further investigation into the TOPIC by DISCIPLINE and TOPIC by OBJECTIVE interactions is warranted. The results of this second analysis indicate the same two statistically significant interactions— TOPIC by DISCIPLINE ($F = 14.10$, df = 118, $p < 0.01$) and TOPIC by OBJECTIVE ($F = 6.46$, df = 59, $p < 0.01$). In addition, statistically significant differences ($p < 0.05$) exist for the interations of OBJECTIVE by INSTITUTION and OBJECTIVE by TOPIC by DISCIPLINE. However, due to the response rates for the three levels of INSTITUTION, as shown in Table 1, conclusions using this variable would be suspect.

5.   Tests are based on Student's t-values.

6.   Tests are based on Student's t-values.

7.   On all multiple-comparison analyses, both the Tukey's Studentized Range (HSD) test and the more conservative Bonferroni (Dunn) T-test were performed. The results of the Bonferroni Test are consistent with the results reported in Table 4 using Tukey's Studentized Range (HSD) test. The Bonferroni test results are available, on request, from the authors.

8.   Statistically significant differences in the magnitude of the CK premium for other topics are present, but none show as clear a pattern as those reported here.

REFERENCES

Accounting Education Change Commission (AECC). 1992. The first course in accounting: Position Statement No. Two. *Issues in Accounting Education* (Fall): 249-251.

Bloom, B.S., ed. 1956. *Taxonomy of educational objectives The classification of educational goals Handbook I: Cognitive domain.* New York: Longman Inc.

Boyer, E.L. 1992. Scholarship reconsidered: Priorities for the professoriate. *Issues in Accounting Education* (Spring): 87-91.

California Society of CPAs. 1995. *California Core Competency Model for the First Course in Accounting* (July).

Cherry, A.A., and S.M. Mintz. 1996. The objectives and design of the first course in accounting from the perspective of nonaccounting faculty. *Accounting Education: A Journal of Theory, Practice and Research,* 1(2): 99-111.

Cherry A.A., and P.M. Reckers. 1983. The introductory financial accounting course: Its role in the curriculum for accounting majors. *Journal of Accounting Education* (Spring): 71-82.

Flaherty, R.E. 1979. *The core of the curriculum for accounting majors.* Sarasota, FL: American Accounting Association.

Frederickson, J.R., and J. Pratt. 1995. A model of the accounting education process. *Issues in Accounting Education* (Fall): 229-246.

Mintz, S.M., and A.A. Cherry. 1993. The introductory accounting courses: Educating majors and onmajors. *Journal of Education For Business* (May/June): 276-280.

Oppenheim, A.N. 1966. *Questionnaire Design and Attitude Measurement.* New York: Basic Books, Inc.

University of Toledo Undergraduate Studies Committee. "Report On Core Curriculum." 1994. Unpublished report.

PROBLEM-SOLVING STYLE AND SUCCESS IN ACCOUNTING CURRICULA

John Sweeney, Carel Wolk, Scott Summers, and
Jim Kurtenbach

ABSTRACT

This study contributes to the extant accounting literature by examining the problem-solving styles of a large sample of accounting majors from two universities and in empirically assessing the relationship between problem-solving style and success in undergraduate accounting curricula. The cognitive problem-solving styles of 222 upper-level accounting students were measured with the Kirton Adaption-Innovation Inventory. Results indicate that accounting students are significantly less innovative than the general population. Additionally, students whose preferred problem-solving style reflected high efficiency had greater success in their accounting curricula, as did students who preferred to solve problems through novel approaches.

Advances in Accounting Education, Volume 2, pages 219-234.
Copyright © 2000 by JAI Press Inc.
All rights of reproduction in any form reserved.
ISBN: 0-7623-0515-0

INTRODUCTION

The AICPA (1998, 1), in its recent *Vision Project*, contends that the future success of the public accounting profession is dependent upon the ability of its members to "become innovative leaders in change." Individuals capable of solving unstructured problems using creative and innovative methods are essential in the complex environment of public accounting (AECC 1990; Arthur Andersen & Co. et al. 1989; Williams 1991; Wolk and Cates 1994; Wolk and Nikolai 1997) and will be in even greater demand in the future (AICPA 1998). Given the dynamic nature of accounting practice today and in the future, it is vital to the success of the profession that new entrants are cognitively capable of innovation and "making sense of a complex and changing world" (AICPA 1998, 2). In this regard, a cognitive characteristic of prime importance is the problem-solving style of new entrants, especially with regards to innovation and dealing effectively with change. A review of the accounting literature, however, reveals a general lack of research concerning the performance effects of different problem-solving styles that accounting majors bring to the classroom and that new staff members bring to the profession. The objective of this study is to contribute to the extant body of empirical research by using a well-established measure from cognitive psychology to examine the relationship between problem-solving style and success in accounting curricula.

Our results indicate that four-year accounting programs may attract students who have problem-solving styles significantly less innovative than the general population. Although the accounting profession is demanding students who work effectively in unstructured settings and thrive in dynamic environments (AICPA 1998), only 29% of our sample of 222 undergraduate accounting majors was classified as innovative problem-solvers. Of the three dimensions of problem-solving style identified by Kirton (1976, 1994), two were significantly associated with success in undergraduate accounting subjects. A preference for efficiency in problem-solving style was associated with a higher accounting GPA, as was a preference for solving problems with novel approaches. The third dimension of problem-solving style, the preference for proliferating original ideas, had no impact on accounting GPA. These findings have pedagogical implications for accounting academicians by highlighting the importance of understanding students' preferred problem-solving styles in creating effective instructional methodologies. The results of this study should also be of interest to administrators of contemporary accounting programs, many of which are implementing changes designed to attract more innovative individuals.

THEORETICAL DEVELOPMENT

The traditional image of an accountant centered on an individual who was efficient, attendant to details, preferred a great deal of structure, and was uncomfortable with change (Bougen 1994). However, this characterization is antithetic to the profession's demand that those entering its ranks work effectively and creatively in diverse settings, be capable of solving unstructured problems, and be leaders in change (AICPA 1998; Wolk and Nikolai 1997). Accounting educators face many obstacles in attracting such dynamic individuals to accounting programs, including restructuring the content and delivery of accounting courses, overcoming long-held stereotypes, competition from other academic programs, and resistance to change from accounting educators (Wolk et al. 1997).

Cultivating a diverse pool of students, especially those capable of innovation and with the ability to function effectively in unstructured situations requires understanding differences in problem-solving styles among individuals. Adaption-Innovation Theory (Kirton 1976, 1994), an area of organizational psychology that is applicable to research on problem-solving, can provide important insights into individual problem-solving styles and how those styles influence success in accounting curricula (Arunachalam et al. 1997).

A key assumption underlying Adaption-Innovation Theory (A-I theory) is that individuals possess, to a greater or lesser extent, one of two preferred cognitive problem-solving styles. These are referred to as the "adaptor" and the "innovator" style. A second assumption is that cognitive style is related to numerous personality traits which appear early in life and which are stable over time (Kirton 1994). Some of these traits are compared in Table 1. A third assumption is that an individual's preferred *style* (approach) is distinct from his or her *level* (capacity or ability) of problem identification and solution (Kirton et al. 1991). Fourth, the theory is essentially "value free" in that one style is not seen as preferable to another, and either style may be the more useful depending on circumstances (Kirton 1976, 1994).

Michael Kirton developed A-I theory in the mid-1970s as a result of his research on organizational change. He theorized, and subsequent research has supported, that individuals prefer either an adaptive problem-solving style, typified by "doing things better" or an innovative style, characterized by "doing things differently" (Kirton 1980, 214). The two styles are reflective of contrasting strengths and weaknesses.

When faced with a problem, adaptors prefer the solution which has the least impact on the assumptions, procedures, and values of the organization; they seek to improve the existing framework and use methods which are safe, secure, and predictable (i.e., they do not "rock the boat"). Adaptors are characterized as being methodical, efficient, reliable and sensitive to group cohesion and cooperation. Although they are essential for the day-to-day functioning of the organization, adaptors have difficulty moving out of established roles in times of change (Kirton

Table 1. Problem-solving Style of Adaptors and Innovators

	Efficiency (E)	Sufficiency vs. Proliferation of Originality (SO)	Conformity to Groups and Rules (R)
ADAPTORS	Concerned with precision, reliability and efficiency	Produces few ideas, generally aimed at improving existing system	Values conformity, stability, consensus and group cohesion
	Capable of doing routine work for long periods	Is concerned with resolving problems, rather than finding them	Is an authority within structured situations
	Seen as disciplined and dependable	Take few risks; resistant to change	Seeks to resolve problems with tried and accepted means
INNOVATORS	Often imprecise, unreliable and inefficient	Produces numerous ideas, some of which appear unsound, generally aimed at changing the existing system.	Is catalyst to settled groups; irreverent of consensus, custom and group norms
	Has little tolerance for routine and structure	Approaches tasks from unusual angles	Takes control in unstructured situations
	Seen as undisciplined and erratic	Change agents; provides the dynamics to bring about periodic radical change	Treats accepted means with little regard; often challenges rules

Source: Adapted from Kirton (1976).

1976). Adaptors tend to produce fewer ideas than innovators, and their ideas or solutions are generally consistent with prevailing practices. An important focus of A-I theory is on the degree of structure preferred by an individual when problem-solving or thinking in general. Adaptors favor well established, structured situations and prefer being consistent with the prevailing reference system (Holland et al. 1991). Audit and tax compliance services, with guiding frameworks of rules or standards and an emphasis on compliance, represent the type of relatively structured occupations preferred by adaptors (Gul 1986; Summers et al. 1998).

In contrast to their more adaptive counterparts, innovators prefer relatively unstructured environments and are at their best when challenged by unique problems requiring novel solutions. Innovators excel at producing new ideas, often cutting across existing paradigms, and are catalysts for organizational change. However, while innovators are essential in times of change, they may have difficulties in applying themselves to standardized tasks and ongoing organizational demands (Kirton 1976, 1994). Innovators can be insensitive to the traditional boundaries of the prevailing paradigm, challenge established methods and customs, and often bring about the radical changes "without which institutions tend to ossify" (Kirton 1976, 623). Consulting services, which lack the structure imposed by formal standards and often require new approaches in resolving unique problems, are expected to be especially attractive to more innovative public accountants (Kirton 1994; Summers et al. 1998).

Kirton (1976, 1994) identified three separate dimensions of problem-solving style that collectively provide an overall characterization of an individual, to a greater or lesser extent, as an adaptor or innovator. The first of these is the *Efficiency* (*E*) component which refers to an individual's preference for thoroughness and attention to detail. Adaptors prefer thoroughness, are concerned with precision, reliability and efficiency, and are capable of performing detailed work for long periods. Innovators do not prefer efficiency, are often viewed as undisciplined, and tend to start many things without following through on them. Innovators become quickly bored with detailed work and tend to delegate routine tasks.

The second component of A-I theory is the *Sufficiency and Originality* (*SO*) dimension, which refers to an individual's preference for idea production and problem solutions. Adaptors prefer to produce a limited number of plausible ideas--limited to the number perceived adequate to address the immediate problem at hand. The ideas produced by adaptors are generally aimed at improving the existing system. Innovators prefer to produce abundant ideas or problem solutions, whether or not they are needed, or even plausible. The ideas produced by innovators are generally aimed at changing the existing system.

The third component is the *Conformity to Rules* (*R*) construct, which refers to an individual's preference for adhering to the prevailing paradigm (its rules, structures, and consensus). Adaptors prefer conformity and are more averse to the risks attendant with shifts in the established way of doing things. In a group setting, adaptors are cooperative, sensitive to the opinions of others and value consensus. Innovators are change agents, less risk averse, and prefer to resist conformity, "valuing more highly the development of their ideas" (Kirton 1987, 18). In a group, innovators often appear insensitive and do "not need consensus to maintain certitude in the face of opposition" (Kirton 1994, 138).

Goldsmith (1984) notes that much of the prior research on creativity and problem-solving has had mixed results because no differentiation was made between "level" and "style." A-I theory is focused only on problem-solving style. It is concerned with an individual's preferred approach to problem-solving as compared to his or her capacity (ability or intelligence) for solving problems (Foxall and Hackett 1992; Kirton 1994). As a result, A-I theory is nonpejorative in that one style of problem-solving is not, a priori, interpreted as better or worse than the other (Kirton 1976, 1994). The demands of the immediate situation will determine the most appropriate problem-solving style. This is in contrast to a level construct, such as intelligence, where higher (i.e., IQ) is seen as "better" than lower.

Prior research has found that innovators are more likely to choose occupations characterized by less structure and with an emphasis on creativity and change. In contrast, adaptors gravitate toward relatively more predictable and structured occupations (Kirton 1994). Summers et al. (1998) found that auditors in public accounting were predominately adaptors while consultants were predominately innovators. Studies by Holland (1987) and Gryskiewicz et al. (1987) indicated that bank employees tend to be adaptors. Hayward and Everett (1983) found that local government employees were predominately adaptors. Keller and Holland (1978) and Kirton and Pender (1982) indicated that R&D personnel tend to be innovators. Managers who work in a more stable environment are inclined to be adaptive problem-solvers, while those who work in more turbulent environments tend toward innovation (Kirton 1994; Thomson 1980).

Research indicates (Kirton 1987, 1994) that a person's cognitive problem-solving style is formed during childhood, changes little over time and is not readily alterable through training. The demands of a particular job or situation, however, may cause an incongruence between preferred problem-solving style and actual behavior. "Coping behavior" results when such circumstances force individuals to operate in a manner that differs from their preferred style (Kirton and McCarthy 1988; Kirton 1994). Prior research has indicated that coping behavior results in increased job stress (Hayward and Everett 1983; Kirton and McCarthy 1988; Lindsey 1985; Summers et al. 1998; Thomson 1985). Kirton (1994, 30) notes that an "excessive demand for coping behavior will not cause a change in cognitive style" but will cause stress and dissatisfaction "give(ing) rise to a desire to leave the situation."

An accounting professional whose preferred problem-solving style is cognitively mismatched with the demands of the job will experience stress and produce coping behavior as a temporary response (Summers et al. 1998). Research with accounting professionals indicates that prolonged stress is likely to result in negative consequences, such as dissatisfaction, reduced commitment and job performance, and a higher intent to turnover (Bamber et al. 1989; Collins and Killough 1992; Choo 1986; Rebele and Michaels 1990). In a study of public accountants, Summers at al. (1998) found that consultants who were classified as innovators and auditors who were classified as adaptors experienced less role stress, greater job satisfaction and organizational commitment, and had lower turnover intentions than consultants who were classified as adaptors and auditors who were classified as innovators. By matching individual problem-solving styles with the demands of the job, public accounting firms may be able to improve performance while reducing stress and its attendant dysfunctional behaviors (Kirton 1994).

In an educational context, accounting students will perform best when the demands of the program are congruent with their preferred problem-solving style (Foxall and Payne 1989; Kirton 1980, 1994). The demands of the accounting program are likely to reflect the problem-solving styles of the instructors. Wolk et al. (1997) found that accounting educators were more likely to be characterized as adaptors (59%) than innovators (41%). Furthermore, adaptive accounting instructors indicated a preference for highly structured classes and a lecture/demonstration presentation mode. Innovative accounting instructors preferred to experiment with a variety of teaching approaches.

Research has characterized accounting students as predominately adaptive problem-solvers. Gul (1986) found more adaptors than innovators in a study of Australian accounting students and Wolk and Cates (1994) found that accounting students were significantly more adaptive in their problem-solving style than were other business majors. Both studies, however, utilized relatively small samples from only one university. Gul (1986) also found a relationship between problem-solving style and career preferences. Adaptors were interested in careers in auditing and banking, while innovators were interested in positions in management consulting. Arunachalam et al. (1997) found that accounting students classified as innovators performed significantly better on unstructured accounting tasks than students classified as adaptors.

HYPOTHESES

In recent years, undergraduate accounting programs have initiated significant changes and pedagogical experimentation in response to changes in the workplace and in technology. However, undergraduate accounting education is still often dominated by structured, textbook-based, rule intensive instructional methodologies (AAA 1986; Arthur Andersen & Co. et al. 1989; AECC 1990; Wolk et al. 1997). This traditional approach to accounting education is consistent with the cognitive style preferences of adaptors and should therefore be more attractive to adaptive students than innovative students. The traditional image of accountants as "bean-counters" wearing "green eye-shades" may also shape images that attract or repel students. Prior research has indicated that the majority of accounting students are adaptors (Gul 1986; Wolk and Cates 1994) and we expect our sample to also reflect an adaptive problem-solving style.

H1. The predominant cognitive problem-solving style of accounting students is adaptive.

Given the relatively structured nature of more traditional four-year undergraduate accounting programs and the rule-based nature of accounting studies in general (AAA 1986; Wolk et al. 1997), it may be expected that within accounting curricula, the adaptive cognitive style will generally result in a comparative advantage over the innovative cognitive style. If A-I theory is to have utility in understanding accounting student performance, however, it will likely need to focus on the three components rather than global comparisons of "adaptors" versus "innovators." Characteristic problem-solving approaches in two of the three components comprising the adaptor-innovator construct, efficiency (E) and conformity to rules and norms (R), would appear to favor the adaptive problem-solvers. The greater efficiency of adaptors should benefit them in a curriculum characterized by structure and methodicalness (Williams 1991). The preference of adaptors for resolving problems within rules and existing frameworks (R) would be advantageous in a structured and rule-bound (i.e., GAAP, GAAS, tax codes, etc.) curriculum. Innovative problem-solvers, however, should have an advantage over their more adaptive counterparts in their capacity to proliferate a quantity of ideas and problem solutions (SO). The following hypotheses regarding relationships between the adaptor-innovator dimensions and success in accounting curricula are proposed:

H2a. Greater efficiency in one's problem-solving style is positively related to success in accounting studies.

H2b. Greater conformity to rules and norms in one's problem-solving style is positively related to success in accounting studies.

H2c. Greater propensity to proliferate original ideas and solutions are positively related to success in accounting studies.

RESEARCH METHOD AND RESULTS

Measures

The Kirton Adaption-Innovation Inventory (KAI) was used to measure individual problem-solving style. It consists of 32 items and utilizes 5-point scales. For each item, respondents indicate how hard or easy ("very hard" to "very easy") it would be for them to maintain a specified innovative or adaptive behavior. KAI scores can theoretically range from 32-160. Observed scores, however, generally range between 45 and 145, and are normally distributed about a mean of 95.3. Studies in more than a dozen countries using the KAI have resulted in similar means (Kirton 1984, 1994). While problem-solving style is a continuous variable, it is common practice in A-I research to dichotomize the sample and compare persons below the mean ("adaptors") with those above the mean ("innovators") (Gul 1986; Foxall 1986; Kirton et al. 1991).

The KAI also computes a score for each of the three dimensions (E, R, and SO) representing the different components of problem-solving style. The E subscale consists of seven items, with lower scores indicating greater efficiency and a more adaptive problem-solving style. The R subscale consists of twelve items, with lower scores (more adaptive) indicating greater conformity to group norms and consensus. The SO subscale contains thirteen items, with lower scores (more adaptive) indicating a preference for producing fewer ideas generally directed at improving the existing system, as opposed to creating a greater number of novel solutions often outside of existing paradigms (more innovative). The additive combination of the three subscale scores equals the total KAI score.

The KAI has been shown to be highly reliable, with Cronbach's alpha consistently ranging in the high .80s across multiple large samples and test-retest coefficients ranging from .82 to .91 (Goldsmith and Matherly 1986; Kirton 1994). The three KAI subscales have also been reported as reliable, with Cronbach's alpha ranging from .70 to .93 and test-retest coefficients generally around .80 (Murdock et al. 1993). Researchers have reported the KAI as having adequate convergent and discriminant validity (Kirton 1987, 1994; Murdock et al. 1993). The KAI is generally uncorrelated with tests of intelligence and levels of education (Goldsmith 1984; Kirton 1976, 1994). Foxall and Hackett (1992) found that the KAI does not confound problem-solving ability with style. Factor analysis (Foxall and Hackett 1992; Taylor 1989) has confirmed that the KAI loads on the three factor structure (*SO*, *E*, and *R*). KAI scores are generally not significantly correlated with measures of social desirability (Goldsmith and Matherly 1986, 1987; Elder and Johnson 1989).

Grade point averages (GPA) in accounting and nonaccounting courses were collected from university transcripts. GPA did not include the accounting course in which data was collected. Descriptive data, including gender, age, and year in program, were collected at the time the KAI was administered.

Subjects

The KAI was administered to 266 junior, senior, and graduate level accounting students in Auditing, Intermediate Financial Accounting, and Accounting Information Systems undergraduate courses at two large Midwestern universities. Both universities had very "traditional," four-year accounting programs and had yet to incorporate changes leading to the 150-hour curriculum. We were unable to obtain GPA data for 35 subjects and these observations were dropped from the sample. One subject did not indicate gender and eight subjects did not indicate their age, resulting in a complete sample of 222.

Table 2 presents sample demographics along with significance levels from ANOVA or pooled *t*-statistics. Participants in the study ranged in age from 19 to 46 years, with an average of 22 years. Fifty-one percent were female; 49% were male. Forty-two percent were juniors, 55% seniors, 3% graduate students. Overall GPA of participants ranged from 1.83 to 4.0 (with a mean of 3.17); GPA in accounting courses ranged from 1.33 to 4.0 (with a mean of 3.08). The average number of completed hours in accounting courses for participating subjects was 16.4 (s.d. = 6).

Female participants were significantly ($p < .001$) more adaptive than their male counterparts (panel A), although both genders attained average KAI scores below the population mean of approximately 95. This result is consistent with prior research, which has found that female accountants (Summers et al. 1998), female

Table 2. Descriptive Statistics by Category

Panel A

Gender	Female n = 113		Male n = 109		T	p	Total Sample	
	Mean	Std Dev	Mean	Std Dev			Mean	Std Dev
KAI	85.09	12.50	92.66	12.52	20.33	0.000	88.81	13.05
E	14.65	3.76	15.71	4.20	3.86	0.051	15.17	4.01
SO	40.55	6.43	44.63	6.81	21.13	0.000	42.55	6.91
R	29.88	6.18	32.32	6.15	8.66	0.004	31.08	6.27
AGE	21.82	3.88	22.63	3.54	2.63	0.106	22.22	3.73
ACCGPA	3.12	0.65	3.03	0.64	0.94	0.334	3.08	0.65
NONGPA	3.24	0.51	3.11	0.50	3.59	0.060	3.17	0.51

Panel B

Class	Audit n = 94		Systems n = 61		Intermediate n = 67		F	p
	Mean	Std Dev	Mean	Std Dev	Mean	Std Dev		
KAI	89.71	14.09	88.41	13.34	87.90	11.23	0.42	0.660
E	14.44	4.16	16.39	4.14	15.09	3.42	4.57	0.011
SO	44.26	7.52	40.84	5.93	41.73	6.39	5.41	0.005
R	31.02	6.63	31.18	6.26	31.07	5.85	0.01	0.988
AGE	22.21	2.19	21.43	3.30	22.96	5.37	2.72	0.068
ACCGPA	3.08	0.65	3.04	0.69	3.11	0.60	0.18	0.833
NONGPA	3.26	0.45	3.18	0.47	3.04	0.59	3.74	0.025

Panel C

Year	Junior n = 93		Senior n = 123		Masters n = 6		F	p
	Mean	Std Dev	Mean	Std Dev	Mean	Std Dev		
KAI	87.31	11.38	89.64	13.03	94.83	29.97	1.51	0.223
E	15.61	3.54	14.72	4.03	17.50	8.36	2.37	0.096
SO	40.78	6.03	43.83	7.02	43.83	11.91	5.45	0.005
R	30.91	5.87	31.09	6.36	33.50	10.46	0.48	0.621
AGE	21.92	4.46	22.26	2.93	26.00	4.73	3.45	0.033
ACCGPA	3.11	0.64	3.02	0.65	3.77	0.30	4.17	0.017
NONGPA	3.14	0.55	3.18	0.47	3.53	0.51	1.63	0.199

Panel D

Major	Acc Undergrad n = 202		Acc Grad n = 11		Nonaccounting n = 9		F	p
	Mean	Std Dev	Mean	Std Dev	Mean	Std Dev		
KAI	88.19	12.58	92.55	9.62	98.00	22.02	2.96	0.054
E	15.10	3.86	15.82	4.45	15.89	6.66	0.31	0.731
SO	42.12	6.86	45.09	3.81	49.22	7.51	5.55	0.004
R	30.97	6.18	31.64	4.46	32.89	9.85	0.45	0.641
AGE	22.26	3.89	21.27	1.19	22.44	1.51	0.38	0.683
ACCGPA	3.07	0.65	3.22	0.68	3.09	0.64	0.27	0.764
NONGPA	3.17	0.51	3.18	0.48	3.24	0.44	0.08	0.924

(continued)

Table 2   (Continued)

Panel E

Univ	University A n = 155		University B n = 67		T	p
	Mean	Std Dev	Mean	Std Dev		
KAI	89.20	13.77	87.90	11.23	0.47	0.495
E	15.21	4.25	15.09	3.42	0.04	0.842
SO	42.91	7.12	41.73	6.39	1.36	0.245
R	31.08	6.46	31.07	5.85	0.00	0.992
AGE	21.90	2.70	22.96	5.37	3.76	0.054
ACCGPA	3.06	0.67	3.11	0.60	0.22	0.640
NONGPA	3.23	0.46	3.04	0.59	6.47	0.012

accounting faculty (Wolk et al. 1997), female accounting students (Wolk and Cates 1994), and females in the general population (Kirton 1994) were, on average, more adaptive than males their male counterparts. Accounting majors, both at the graduate and undergraduate level (panel D), had lower KAI scores ($p < .054$) than participants who were not accounting majors, although the small number of nonaccounting majors ($n = 9$) in the subject pool makes comparisons precarious.

Students from both universities represented in the study were very similar, based on the variables reported in panel E. No significant differences were noted in any of the KAI measures or in accounting GPA. Students from University A were, on average, younger by approximately one year and had slightly higher GPAs in nonaccounting courses. KAI scores by class (panel B) and by year in college (panel C) demonstrate little variation, although the *E* and *SO* subscales do vary significantly across these categories. We cannot proffer a reasonable explanation as to why the *E* and *SO* subscales would vary across class and year. Panel C does indicate that the accounting GPA of junior subjects was higher than that of senior subjects, perhaps due to the difficulty of more advanced accounting courses.

Statistics

As only subscale scores and not raw responses were tabulated in our database, Cronbach's alpha was not computed for the subscales. However, as previously noted, prior research has reported the KAI and its subscales as having adequate internal reliability (Kirton 1994). Participants' KAI scores ranged from 62 to 146, with a mean of 88.1 (s.d. = 13.05). This mean is significantly below the population mean of 95.3 reported by Kirton (1994) ($t = 7.00$, $p < .001$). Consistent with prior research, female participants were, on average, more adaptive than males. Mean female KAI score was 85.1, as compared to the mean male KAI score of 92.7 ($p < .001$). There was no difference in KAI score, subscale scores or accounting GPA between students at the two participating universities. There was a significant difference in average age (21.90 vs. 22.96) and non-accounting GPA (3.23 vs. 3.04) between subjects at the two universities.

As hypothesized (H1), the majority of participants were adaptive problem-solvers. Dichotomizing the sample into adaptors and innovators (≤ 95 = adaptor; > 95 = innovator), indicates that significantly ($p < .01$) more accounting students (158 or 71%) were classified as adaptors than as innovators (64 or 29%). These results are consistent with the Wolk and Cates' (1994) small sample study ($n = 39$) of accounting students, which included 77% adaptors and 23% innovators.

The primary hypotheses of the study, H2a-H2c, examine the relationships among the KAI subscales and success in accounting studies, as measured by GPA in accounting courses (ACCGPA). These hypotheses were tested through the regression model in Table 3. GPA in nonaccounting courses (NON-ACCGPA) was included in the model as a variable to control for individual ability. Sex (GENDER), year in program (junior, senior, masters), major (undergraduate accounting, graduate accounting, nonaccounting major), and age (AGE) are also included as control variables in the model.[1]

The regression model was significant ($p < .0001$) and accounted for 43% of the variance in the dependent variable, ACCGPA. The control variable NON-ACCGPA had a strong relationship with success in accounting studies. As would be expected, subjects who performed well in non-accounting courses also performed well in accounting courses. AGE was significant in the model, with older students obtaining higher grades in their accounting courses. GENDER and major were not influential variables on success in accounting courses. Subjects'

Table 3.   Regression Analysis—ACCGPA
(n = 222)

Source	Sum of Squares	df	Mean Square	F	
Model	42.44	10	4.24	17.85	Sig.
	Beta	Std. Error	t	Sig.	0.000
Intercept	−0.0230	0.443	−0.05	0.959	
E	−0.0280	0.011	−2.63	0.005	
SO	0.0001	0.006	0.02	0.496	
R	0.0205	0.007	2.90	0.002	
NON-ACCGPA	0.7471	0.070	10.72	0.000	
GENDER	−0.0051	0.071	−0.07	0.942	
Senior Year (0/1)	−0.1544	0.070	−2.21	0.028	
Masters Year (0/1)	0.2761	0.215	1.28	0.201	
Acc Grad. Major (0/1)	0.1744	0.153	1.14	0.255	
Non Acc Major (0/1)	−0.0534	0.171	−0.31	0.756	
AGE	0.0266	0.009	2.85	0.005	

Note: R Squared = .458 (Adjusted R Squared = .433)

level in program' (junior, senior, masters) did exert some influence on ACCGPA. Students who were seniors had significantly lower ACCGPA than those at the junior or masters level.

Hypothesis H2a is supported as greater efficiency (low E score—more adaptive) results in a higher GPA in accounting courses. The negative coefficient for the E subscale is consistent with our expectation: students with lower E scores (more efficient) are more successful in accounting courses. Although the R subscale is also significant in the model, the sign of the coefficient is contrary to that hypothesized in H2b. Rule-oriented (low R score—more adaptive) students, who have a preference for solving problems under the framework of well established methods, attained a lower GPA in accounting courses than students who prefer to solve problems with novel and less established approaches (high R score—more innovative). The SO variable (Hypothesis H2c) was not significant in the model. The ability to proliferate a quantity of original ideas (high SO—more innovative) did not have an impact on accounting GPA.

DISCUSSION AND IMPLICATIONS

The recent "CPA Vision Project" (AICPA 1998, 2) stated that the core purpose of CPAs is "Making sense of a changing and complex world." In recognition of the changing nature of accounting services, considerable impetus has been provided by the Accounting Education Change Commission (AECC 1990) and the AICPA (1998) to encourage accounting programs to attract and retain students who will be able to effectively function in a changing and complex world. The dynamic nature of contemporary accounting practice seems to require more accountants with an innovative problem solving style. The majority of participating accounting students in this study and in prior research, however, have been adaptive problem solvers (Gul 1986; Wolk and Cates 1994). Furthermore, as public accounting firms increasingly emphasize less structured consulting services over more structured products, such as auditing and tax preparation services, innovative accountants should be in even greater demand in the future (AICPA 1998; Summers et al. 1998).[2]

The results of this study indicate that the curricula of the undergraduate accounting programs represented in the sample do not appear to consistently favor either the adaptive or the innovative problem-solving style. Greater efficiency (E), an adaptive trait, was associated with higher accounting GPA. However, a preference for resolving problems under the framework of established methods and rules (R), an adaptive approach, resulted in a lower accounting GPA. The innovator's preference for new approaches to resolving problems appears advantageous in undergraduate accounting studies. This result was unexpected, given the rule/standards orientation inherent in many accounting subjects.

A potential concern to accounting educators is the insignificant relationship between the ability to produce original ideas (*SO*) and accounting GPA. The lack of significance may indicate that the capacity to produce a greater number of diverse and original problem solutions, characteristic of the innovative style and demanded by the accounting profession (AICPA 1998; Arthur Andersen & Co. et al. 1989), is not accentuated in the more traditional programs of accounting education. Prior research (Arunachalam et al. 1997) has indicated that innovators outperform adaptors on unstructured accounting tasks. Incorporation of more unstructured assignments emphasizing original and creative thinking may help in attracting more innovative students to the accounting major.

In order to attract the more innovative problem-solvers that the public accounting profession appears to demand (AICPA 1998), accounting faculty may need to make pedagogical changes in the instruction of courses and in the evaluation of students. This metamorphis can be at least partially accomplished by implementing the following recommendations. First, an understanding of the adaptor-innovator construct and subscales will enable accounting faculty members to be aware of the strengths and weaknesses of both styles. Second, faculty members need to make a conscious effort to vary the type and structure of assignments and teaching modalities to reward and reinforce both the adaptive and innovative approaches (Kirton et al. 1991; Arunachalam et al. 1997). Third, given the recent emphasis placed upon cooperative learning (AECC 1990), educators should consider the adaptor-innovator dynamic when forming groups (Wolk et al. 1997). Fourth, faculty members should be aware of their own problem-solving style, and the potential biases their characteristic approach may introduce in the classroom (Wolk et al. 1997). Administrators of accounting programs can also actively recruit students from outside areas who may bring to the discipline a broader range of problem-solving styles. In conjunction with this effort, accounting programs should make a concerted effort to replace the traditional image of the accountant with a more contemporary version (Bougen 1994) appealing to individuals with a broader range of problem-solving styles.

These findings have implications for future research and for accounting programs attempting to attract students capable of meeting the demands of a changing profession. Future research should be directed towards investigating how differences in course content and instructional methodologies interact with individual problem-solving styles, particularly at the subscale level, in affecting students' performance on specific competencies. Additionally, understanding how course content and structure influences students' self-selection processes into accounting courses may be particularly important to programs redesigning the traditional four-year accounting curricula.

It is important to recognize the limitations of this research effort. First, generalizability may be limited as subjects represented only two universities and were not chosen randomly from geographically diverse accounting programs. Second, the focus of this study was on overall performance/GPA in accounting courses, which is an aggregate measure. Potential employers of accounting students, however, may be more interested in specific competencies, such as in systems, auditing, taxation, etc. The use of an aggregate performance measure does not reflect specific strengths and weaknesses. Third, the reliability of the study is limited by potential error in the KAI's measurement of the three subscale constructs. Fourth, the relationship between performance in accounting courses and problem-solving style applies only in an academic setting and does not necessarily imply a relationship between accounting job performance and problem-solving style.

REFERENCES

Accounting Education Change Commission (AECC). 1990. *Position Statement Number One: Objectives of Education for Accountants* (September).

American Accounting Association, Committee on the Future Structure, Content, and Scope of Accounting Education (AAA). 1986. Future accounting education: Preparing for the expanding profession. *Issues in Accounting Education* (Spring): 168-195.

American Institute of Certified Public Accountants (AICPA). 1998. *CPA Vision Project*. New York: AICPA.

Arthur Anderson & Co, Arthur Young, Coopers & Lybrand, Deloitte, Haskins & Sells, Ernst & Whinney, Price Waterhouse, & Touche Ross. 1989. *Perspectives on Education: Capabilities for Success in the Accounting Profession.*

Arunachalam, V., J. Kurtenbach, J., and J. Sweeney. 1997. The relationship between cognitive problem-solving style and task structure in affecting student performance. *The Accounting Educators Journal* (Fall): 1-13.

Bamber, E., D. Snowball, and R. Tubbs. 1989. Audit structure and its relations to role conflict and role ambiguity: An empirical investigation. *The Accounting Review* (April): 285-299.

Bougen, P.D. 1994. Joking apart: The serious side to the accountant stereotype. *Accounting, Organizations and Society* 19: 319-335.

Choo, F. 1986. Job stress, job performance, and auditor characteristics. *Auditing: A Journal of Practice and Theory* (Spring): 17-34.

Collins, K., and L. Killough. 1992. An empirical examination of stress in public accounting. *Accounting, Organizations and Society* 17: 535-547.

Elder, E.L., and D. C. Johnson. 1989. Vary relationships between Adaption-Innovation and social desirability. *Psychological Reports* 70: 169-170.

Foxall, G.R. 1986. Managerial orientations of adaptors and innovators. *Journal of Managerial Psychology*, 24-27.

Foxall, G.R., and P.M. Hackett. 1992. The factor structure and construct validity of the Kirton Adaption-Innovation Inventory. *Personality and Individual Differences* 13: 967-975.

Foxall, G.R., and A.F. Payne. 1989. Adaptors and innovators in organizations. A cross-cultural study of the cognitive styles of managerial functions and sub-function. *Journal of Human Relations* 40(7): 639-650.

Goldsmith, R.E. 1984. Personality characteristics associated with adaption-innovation. *The Journal of Psychology*, 159-165.

Goldsmith, R.E., and T.A. Matherly. 1987. Adoption-innovation and creativity: A replication and extension. *British Journal of Social Psychology* 26: 79-82.

Foxall, G.R., and A.F. Payne. 1987. Adaption-innovation and creativity: A replication and extension. *British Journal of Social Psychology* 26: 79-82.

Gryskiewicz, S.S., D.W. Hills, Holt, and K. Hills. 1987. *Understanding Managerial Creativity: The Kirton Adaption-Innovation Inventory and Other Assessment Measures.* Greensboro, NC: Center for Creative Leadership.

Gul, F.A. 1986. Adaption-Innovation as a factor in Australian accounting undergraduates' subject interests and career preferences. *Journal of Accounting Education* (Spring): 203-209.

Hayward, G., and C. Everett. 1983. Adaptors and innovators: Data from the Kirton Adaption-Innovation inventory in a local authority setting. *Journal of Occupational Psychology* 56: 339-342.

Holland, P.A. 1987. Adaptors and innovators: Application of the Kirton Adaption-Innovation inventory to bank employees. *Psychological Reports* 60: 263-270.

Holland, P.A., I. Bowskill, and A. Bailey. 1991. Adaptors and innovators: Selection versus induction. *Psychological Reports* 68: 1283-1290.

Keller, R.T., and W.E. Holland. 1978. Individual characteristics of innovativeness and communication in research and development organizations. *Journal of Applied Psychology*, 759-762.

Kirton, M.J. 1976. Adaptors and innovators: A description and measure. *Journal of Applied Psychology*, 622-629.

Kirton, M.J. 1980. Adaptors and innovators in organizations. *Human Relations*, 213-224.

Kirton, M.J. 1987. *Kirton Adaption-Innovation Inventory Manual*, 2nd ed. Hatfield, UK: Occupational Research Centre.

Kirton, M.J. 1994. *Adaptors and Innovators: Styles of Creativity and Problem Solving.* London: Routledge.

Kirton, M.J., A. Bailey, and W. Glendinning. 1991. Adaptors and innovators: Preference for educational procedures. *The Journal of Psychology*, 445-455.

Kirton, M.J., and R.M. McCarthy. 1988. Cognitive climate and organizations. *Journal of Occupational Psychology* 61: 175-184.

Kirton, M.J., and S.R. Pender. 1982. The adaption-innovation continuum: occupational type and course selection. *Psychological Reports* 51: 883-886.

Lindsey, P. 1985. Counseling to resolve a clash of cognitive styles. *Technovation*, 57-67.

Murdock, M., S. Isaksen, and K. Lauer. 1993. Creativity training and the stability and internal consistency of the Kirton Adaption-Innovation Inventory. *Psychological Reports* 72: 1123-1130.

Rebele, J.E., and R.E. Michaels. 1990. Independent auditors' role stress: Antecedent, outcome, and moderating variables. *Behavioral Research in Accounting* 2: 124-153.

Summers, S.L., J.T. Sweeney, and C.M. Wolk. 1998. The relationship between cognitive problem solving style, stress and outcomes in public accounting: Consulting vs. audit. Presented at the 1998 Western Regional Meeting and the 1998 Annual Meeting of the American Accounting Association.

Taylor, W.G.K. 1989. The KAI: A re-examination of the factor structure. *Journal of Organizational Behaviour* 10: 297-307.

Thomson, D. 1980. Adaptors and innovators: A replication study on managers in Singapore and Malaysia, *Psychological Reports* 47: 383-387.

Williams, D.Z. 1991. The challenge of change in accounting education. *Issues in Accounting Education* (Spring): 126-133.

Wolk, C.M., and T.A. Cates. 1994. Problem-solving styles of accounting students: Are expectations of innovation reasonable? *Journal of Accounting Education* (Fall): 269-281.

Wolk, C.M., and L.A. Nikolai. 1997. Personality types of accounting students and faculty: comparisons and implications. *Journal of Accounting Education* (Winter): 1-17.

Wolk, C.M., T. Schmidt, and J. Sweeney. 1997. Accounting educators' problem-solving style and their pedagogical perceptions and preferences. *Journal of Accounting Education* (Fall): 469-483.

THE ETHICS CONSTRUCT:
A MULTIDIMENSIONAL ANALYSIS
IN AN ACADEMIC SETTING

Daryl M. Guffey and Mark W. McCartney

ABSTRACT

This study had two stated purposes. One was to test a multidimensional measure of ethical behavior in a new setting with different experimental subjects. The objective was to test the measure's generalizability in gauging ethical decision making. The other purpose was to draw inferences on three cognitive dimensions of ethical judgment for pedagogical purposes, concluding with a discussion of implications for teaching ethics in an accounting setting. Based on responses from 268 subjects at seven large public universities, we found the dimensions of moral equity, relativism, and contractualism to capture the construct of ethical judgment. Tests also indicated that these three dimensions explained a substantial portion of the variance in the determination of whether or not an action is judged to be ethical. These dimensions also captured a substantial amount of the variance in the construct "behavioral intention."

Advances in Accounting Education, Volume 2, pages 235-256.
ISBN: 0-7623-0515-0

INTRODUCTION

The practice of public accounting depends on ethical behavior. Because of the public's faith in the integrity of public accountants, certified public accountants (CPAs) regularly receive confidential information. Also, the public believes that audits add value to financial statements because CPAs are competent, diligent and perform objective examinations. To underscore the importance of ethics to the profession, the American Institute of Certified Public Accountants (AICPA) details an ethical code to which its members must adhere.

A code of conduct is an important step in fostering ethical behavior in the accounting profession. However, given the importance of ethics to the profession, an adequate understanding of ethical decision making and its influential factors is also desirable. This understanding will hopefully lead to mechanisms that, accompanied with a code of conduct, are effective in encouraging ethical conduct and decision making. In fact, a code with its enforcement mechanisms, alone, is probably not enough to effectively encourage ethical conduct. Noreen concludes this, stating that "...an agreement (i.e., ethical code) to abstain from opportunistic behavior cannot be enforced effectively by external rewards or sanctions; instead, the sanctions for unethical behavior must be internalized" (1988, 359). In other words, it would be desirable to instill ethical standards in peoples' belief systems such that the longing to act unethically is diminished. A greater understanding of the dynamics that drive ethical decision making may lead to the development of processes that adequately infuse such standards in one's belief system, creating a greater desire to act ethically. Flory, Phillips, Reidenbach, and Robin (1992, 288) support this, stating that "...internalization of these sanctions calls for a better understanding of ethical behavior and of the multidimensional factors involved in shaping ethical conduct."

This study had two purposes. First, we tested a multidimensional ethics measure in an academic setting. Reidenbach and Robin (1990) developed the measure, positing that three cognitive dimensions—moral equity, relativism, and contractualism—operate together to influence ethical decision making. Flory, Phillips, Reidenbach, and Robin (1992) and Robin, Reidenbach, and Forrest (1996) found the three-dimensional scale to reliably capture behavioral intent and decision making among management accountants and advertising managers, respectively, who were exposed to scenarios depicting realistic ethical problems faced by members of their professions. Measures of broad constructs, such as "ethics," should be tested in a variety of settings and with different groups to confirm that initial results are not an artifact of a particular group or setting. We tested the multidimensional measure using upper level accounting students who evaluated four scenarios involving questionable academic practices. Accounting students operate under a somewhat different set of demands than do accounting and advertising professionals. If the results of this study support those of Flory, Phillips, Reidenbach, and Robin (1992) and Robin, Reidenbach, and Forrest

(1996), additional evidence will exist that the multidimensional measure exhibits a high level of construct validity. More evidence will also exist that the same factors are significant in influencing ethical perceptions over varying ethical problem situations and for individuals operating in different ethical environments.

The second purpose of this study was to gain a better understanding of the ethical decision making processes of undergraduate accounting students. Increased emphasis has been placed on teaching ethics in an accounting curriculum. The report of the Treadway Commission (National Commission on Fraudulent Financial Reporting 1987, 82-83) supports the teaching of ethics in accounting and business curriculums. Smith (1993) surveyed auditing professors and found that an overwhelming majority agreed that ethics and personal integrity should be taught. Other articles highlight academics' concern for effectively teaching accounting ethics. These articles focused on items such as the goals of teaching ethics in accounting, the location of accounting ethics in the curriculum, who should teach accounting ethics, and appropriate techniques for teaching accounting ethics.[1] Northwestern University researchers demonstrated the potential rewards of effectively teaching ethics. They examined the impact of a mandatory ethics program in Chicago's public schools. Where the program was at least partially implemented, 70% of school principals found significant reductions in truancy, and 80% reported substantial declines in disciplinary problems (Nazario 1990). More directly related to the current research, Green and Weber (1997) exposed auditing students to the AICPA Code of Professional Conduct, and found that the students reasoned at a higher level of ethical development than did a control group not exposed to the Code. Given this evidence suggesting that teaching ethics assists in shaping students' ethical processes, a better understanding of the forces that drive ethical decision making among students will lead to increasingly effective methods of teaching accounting ethics.

The next section of this paper presents a review of ethics research in accounting. Then we present a more thorough description of the multidimensional ethics measure. Following that is a presentation of our survey and research methodology. The results are then discussed, which consider the issues of internal consistency as well as content and predictive validity of the measure. The last two sections of this paper provide implications of this research for teaching accounting ethics and our conclusions.

ACCOUNTING LITERATURE

Numerous accounting studies investigated the moral development of accounting students and whether ethics can be taught. Several studies have employed the Defining Issues Test (DIT) to evaluate students' level of moral reasoning. The DIT provides a P (principled) score that indicates the level of moral reasoning. Higher P scores indicate higher levels of moral reasoning.

Jeffrey (1993) found that senior students in each business major had higher levels of ethical development than entering students in each major. This finding suggests that ethics can be taught. However, Ponemon (1993) found that ethics cannot be taught, at least with respect to moral development and free-riding behavior. His results indicated that ethics intervention did not cause accounting students' levels of ethical reasoning to develop, and it did not curtail students' free-riding behavior. Armstrong (1993) recommended teaching ethics using a "sandwich approach." That is, students take a general ethics course outside the business school followed by accounting courses with case studies and homework involving ethical issues. The process concludes with a capstone course in ethics and professionalism. In contrast to Ponemon (1993), Armstrong found a positive effect on the moral development of students.

Ponemon and Glazer (1990) explored the ethical development of students and alumni in public accounting practice from two educational institutions using the DIT. They found that DIT P scores of seniors and alumni from each school, on average, are greater than P scores for freshmen from the same institution, indicating that moral reasoning of accounting students increases during the college years. Second, the P score variation within the alumni strata is significantly lower than variation within student ranks for both institutions. This finding is indicative of socialization within accounting firms and is consistent with results reported by Ponemon (1990). Ponemon and Glazer (1990) also found that students and alumni from the school offering a liberal arts curriculum, on average, were more highly developed in terms of DIT measures than students and alumni of the school with a more traditional accounting program.

Lampe and Finn (1994) collected data from over 300 auditors and combined their data with data from several other cognitive-developmentalist studies. They concluded that alternative strategies for teaching ethics to accounting students lead to differing levels of moral development which, in turn, have a significant impact on the ethical decisions they make. They found that accounting practitioners' DIT P scores were significantly lower than for college graduates. In addition, the average P score for accounting students from traditional business colleges were significantly lower than the average for liberal arts college students.

Numerous accounting studies have explored alternative methods of incorporating ethics education in accounting education. Hiltebeitel and Jones (1991) found that, after completing ethics modules in accounting courses, students revised the manner in which they resolved professional ethical dilemmas. They suggest that, in lieu of trying to teach students "right from wrong," a better approach is to provide students with an effective method for resolving ethical dilemmas. Mintz (1995) presented the meaning and purpose of virtue theory and how it relates to accounting practice. Furthermore, he described how pedagogical approaches such as case analysis, role-playing, and collaborative learning techniques can be used to teach virtue to accounting students. Beets (1993) found that role-playing enhanced student understanding of professional ethics. LaGrone, Welton, and

Davis (1996) provide evidence that ethics education must become part of continuing education programs to be effective. They found that gains from ethics intervention in a graduate accounting course are transitory and should be reinforced through continuing education programs. Green and Weber (1997) exposed auditing students to the AICPA Code of Professional Conduct, and found that the students reasoned at a higher level of ethical development than did a control group not exposed to the Code.

Our paper investigated these issues from a different perspective—the multidimensional ethics measure. Tests for construct validity produced supportive results for the hypothesized three-dimensional measure. We first developed a questionnaire with four scenarios concerning ethical issues. Next, the results from a factor analysis and a reliability coefficient test (coefficient alpha) suggested that a high degree of internal consistency existed for each dimension of the measure. We then tested the content validity of the three-dimensional measure by comparing it with a global ethical/unethical measure. Last, a sense of predictive validity was obtained by comparing the multivariate measure with a behavioral intention measure for the respondents. All the test criteria supported the proposition that the three dimensions—moral equity, relativism, and contractualism—capture a substantial amount of the decision dynamics used by the respondents to make ethical judgments. The next section presents a more thorough description of the multidimensional ethics measure.

MULTIDIMENSIONAL ETHICS MEASURE

Crucial to the development of models of ethical decision making is development of valid and reliable measuring devices. Reidenbach and Robin's (1990) starting point in developing an ethics scale was a content analysis of the contemporary normative moral philosophies. From there, they followed the assumption that individuals use more than one rationale in making ethical judgments, and that the importance of those rationales is a function of the problem situation faced by the individual. Their product was a multidimensional measure, illustrated in Figure 1, consisting of eight bipolar scales divided into three dimensions—moral equity, relativism, and contractualism.

The moral equity dimension is comprised of the following four items: (1) Morally/not morally right, (2) Just/unjust, (3) Fair/unfair, and (4) Acceptable/unacceptable to my family. Reidenbach and Robin (1990, 646) suggest that this dimension relies heavily on lessons from our early training received in the home regarding fairness, right and wrong as communicated through childhood lessons of sharing, religious training, and morals from fairy tales and fables. Morally/not morally right comes from deontology, which suggests that individuals have duties, such as the duty to pay one's debts, care for one's children, and tell the truth because it is the "right" thing to do [Reidenbach and Robin (1990, 651)]. Put

Moral Equity Dimension

Morally Right____ : ____ : ____ : ____ : ____ : ____ : ____Not Morally Right

Just____ : ____ : ____ : ____ : ____ : ____ : ____Unjust

Fair____ : ____ : ____ : ____ : ____ : ____ : ____Unfair

Acceptable to____ : ____ : ____ : ____ : ____ : ____ : ____Not Acceptable to
My Family   My Family

Relativism Dimension

Culturally____ : ____ : ____ : ____ : ____ : ____ : ____Culturally
Acceptable   Unacceptable

Traditionally____ : ____ : ____ : ____ : ____ : ____ : ____Traditionally
Acceptable   Unacceptable

Contractualism Dimension

Violates an____ : ____ : ____ : ____ : ____ : ____ : ____Does not violate an
Unwritten Contract   Unwritten Contract

Contractualism Dimension

Violates an____ : ____ : ____ : ____ : ____ : ____ : ____Does not violate an
Unspoken Promise   Unspoken Promise

Figure 1.   The Multidimensional Ethics Measure

another way, Reidenbach and Robin (1990) used "morality" as defined by Beauchamp (1982, 5), who stated that, "...morality is concerned with many forms of
belief about right and wrong human conduct." They then referred to the moral
development literature of Rest (1979), which relies heavily on the concept of justice in associating fair and just with the moral equity dimension. Rest (1979,
35-36) used the concept of procedural justice, having as its purpose the development of rules or procedures that result in fair or just outcomes. Acceptable/not
acceptable to my family is at its base a relativistic concept, that all normative
beliefs are a function of a culture or individual, and therefore, no universal ethical
rules exist that apply to everyone [Reidenbach and Robin (1990, 651)]. The relativistic argument continues that, since ethical rules are relative to a specific culture (e.g., one's family), the values and behaviors of people in one culture need
not govern the conduct of people in another culture. An explanation of why this
item is an element of the moral equity dimension is included in the following
paragraph's discussion of the relativistic dimension. The moral equity dimension
represents that decisions are evaluated essentially in terms of their inherent
fairness, justice, goodness and rightness.

The relativistic dimension is comprised of the following two items: (1) Traditionally acceptable/unacceptable, and (2) Culturally acceptable/unacceptable. This dimension is more concerned with the guidelines, requirements, and parameters inherent in the social/cultural system than with individual considerations. These beliefs are relativistic in the sense that beliefs are subject to the dictates of society. Reidenbach and Robin (1990, 646) suggest that this dimension is acquired as the individual experiences adequate and sufficient social intercourse to develop greater understanding of cultural and traditional norms. It may be noted that "acceptable/not acceptable to my family" appears to fit the definition of a relativistic item. However, it appears in the moral equity dimension because this cultural norm is associated with the family as opposed to the cultural/societal norms outside of the family.

The third dimension is referred to as the contractualism dimension, and consists of the following two items: (1) Violates/does not violate an unspoken promise, and (2) Violates/does not violate an unwritten contract. Reidenbach and Robin (1990, 646) describe this dimension as resembling most closely the idea of a "social contract" that exists between business and society. They state that most, if not all, business exchanges incorporate either implicit or explicit promises or contracts; a *quid pro quo* wherein one party is obligated to provide a product, service, employment, or perform some action in return for something of value. The view of business exchanges is then broadened beyond a purely economic nature to include notions of fair play, truth telling, duty, and rights.

RESEARCH METHODOLOGY

Three hundred and nine students in upper-level accounting courses participated in the study. The students were enrolled in seven large public universities in the United States, one in the northeast, two in the southeast, two in the southwest, one in the midwest, and one in the west.[2]

Only responses from complete research instruments were used, reducing the sample size to 268 subjects (a response rate of 86%). Survey instruments in which subjects omitted demographic data were included in the study. Ages of the participants ranged from 19 to 68, with the average being 24.6 years. Over 56% (151/268) had completed a course in ethics. Accounting majors comprised almost 89% of the sample (238/268), which consisted of 141 females (54.8%) and 127 males (45.2%). The overall grade point average (GPA) for the sample was approximately 3.13 with 62% (165/268) having a GPA over 3.0 and over 23% (64/268) having a GPA over 3.5. Over three-fourths of the sample were seniors and graduate students. Thus, the majority of students performed well in the classroom and appear ready to assume positions within accounting firms, industry, or government.

Research Instrument

A survey questionnaire was used and students were assured that their participation was voluntary and that their responses would remain anonymous. Participants were first requested to provide basic demographic information such as classification, gender, age, GPA, and major. They were then asked to complete the eight-item scale for four academic cheating scenarios. The scenarios, presented in the appendix to this paper, were each approximately 200 words in length and portrayed realistic questionable ethical academic situations. Each scenario included an action statement to assure that all respondents were reacting to the same stimulus.

The topics for the scenarios came from Ameen, Guffey, and McMillan (1996), who surveyed 320 upper-level accounting students regarding the perceived seriousness of 23 questionable academic practices related to exams, projects, and written assignments. In selecting scenarios, our primary concerns were that they be understandable to the students and that they provide a vigorous validity test of the multidimensional ethics measure. Flory, Phillips, Reidenbach, and Robin (1992, 288-291) state that psychometricians consider validation of a measure of a broad construct, such as "ethics," to be unending. They go on to say that, "Repeated use of the same measurement items...to evaluate different situations...is an important test" of the measure's validity. Therefore, we selected scenarios that varied in their perceived seriousness to the students in the Ameen et al. (1996) study. If the same three dimensions apply as anticipated, other scenarios that significantly differ from one another in terms of seriousness, a stronger test of the applicability (validity) of this instrument as a measure of ethical decision making will have been demonstrated. Two of the scenarios selected for this study were not considered to be particularly serious by the Ameen et al. (1996) subjects. Scenario D, with a mean score of 1.48, described a student not reporting an arithmetic error made by a professor on her exam.[3] The error increased her exam grade by thirty points. Scenario B had a mean of 1.98, and described a student being informed of material on an exam by another student who took that same exam earlier in the day. Scenarios A and C had average scores of 4.11 and 4.14, respectively. Scenario A described a student who purchased a term paper from a commercial research firm and C told of a student who found a copy of an upcoming exam in the trash, and took the copy home for assistance in preparing for that exam. The differences between either of the less and either of the more serious scenarios was significant at $p < 0.01$.

Three types of tests were used to validate this instrument, as advocated by Nunnally (1978, 87). Construct validity was the first to be measured (i.e., that the instrument measures "ethical decision making"). Content validity, an assessment of the adequacy of the scale items in measuring the ethical judgment construct, was then measured. Last, predictive validity, the effectiveness of the measure as a predictor of some criterion or dependent variable, was assessed. The sections that follow describe relevant findings for each of these three types of validity.

RESULTS: TESTS FOR VALIDITY

Ethical judgment is a considerably complex construct. We hypothesize that the dimensions of moral equity, relativism, and contractualism influence the ethical judgments of accounting students; that is, each dimension represents a part of the ethical judgment construct.

Construct Validity

Assessing construct validity is directly concerned with the theoretical relationship of the scores on the measurement instrument with the construct it purports to represent—ethical judgment. That is, does the measure "behave" the way the construct it purports to measure should behave? Construct validity was confirmed by focusing on the reliability (internal consistency) of the scales (test items). Two techniques were used to determine scale reliability: factor analysis and coefficient alpha.

Factor analysis allows one to identify the separate dimensions (factors) being measured by a survey questionnaire. Its purpose, generally speaking, is to condense the information contained in the original variables into a smaller set of new composite dimensions with a minimum loss of information (Hair, Anderson, and Tatham 1987, 235). The survey instrument contained eight original questions. If the dimensions operate as hypothesized, these eight questions should group together naturally when analyzed. That is, three factors should be produced, each with the anticipated scale measures for their respective dimensions.

The subjects in this study were presented with four separate ethical scenarios. Therefore, four factor analyses can be conducted, one for each scenario. To the extent that the scale is reliable, variation in scale scores can be attributed to the true score of the phenomenon being measured, ethical judgment. If, in the four independent factor analyses, the scales produce the same results, they exhibit a high degree of reliability.

The factor analyses were conducted using a principal components analysis with an orthogonal rotation. Therefore, each factor is independent of all other factors and the correlation between factors is arbitrarily determined to be zero.[4] Consistent with Flory et al. (1992, 292) the principal component outcomes were constrained to confirm the three anticipated factors. Table 1 contains the results of the factor analyses. In general, the three anticipated dimensions (moral equity, relativism, and contractualism) are apparent in each of the scenarios. The factor loadings range from 0.47 to 0.91 with only one in the 0.40's, five in the 0.60's, four in the 0.70's, thirteen in the 0.80's, and nine in the

Table 1. Factor Structures for the Four Scenarios

	Factor Loadings											
	Factor 1				Factor 2				Factor 3			
Scenario	A	B	C	D	A	B	C	D	A	B	C	D
Moral Equity Dimension												
Fair/unfair	.88	.89	.90	.83	.10	.17	.17	.37	.13	.24	.14	.19
Just/unjust	.90	.88	.91	.85	.11	.23	.16	.35	.06	.22	.12	.18
Morally right/not morally right	.66	.76	.68	.77	.28	.35	.32	.28	.14	.22	.34	.11
Acceptable/unacceptable to family	.47	.66	.60	.40	.39	.46	.46	.66	.19	.24	.18	.17
Relativism Dimension												
Traditionally acceptable/unacceptable	.06	.23	.20	.38	.89	.91	.89	.83	.07	.11	.02	.11
Culturally acceptable/unacceptable	.32	.32	.24	.27	.73	.86	.86	.89	-.03	.15	.07	.13
Contractualism Dimension												
Violates/does not violate promise	.01	.24	.11	.16	.03	.13	.09	.16	.89	.90	.91	.90
Violates/does not violate contract	.28	.25	.25	.14	.06	.15	.02	.10	.79	.90	.86	.91

Note: The scale item and factor associations are underlined. The percentage of variance explained by the three-factor solution is as follows:
Scenario A, 68.8%;
Scenario B; 84.3%
Scenario C; 78.7%
Scenario D; 81.4%

0.90's. The amount of variance explained by each of the four factor analyses ranged from a low of 68.8% to a high of 84.3%.

Table 1 does indicate some imperfection in the factor loadings. One moral equity item (Acceptable to My Family/Not Acceptable to My Family) loads inconsistently. For scenario D it loads on the relativism dimension, while it loads more heavily on the moral equity factor for scenarios A, B, and C. With this one exception, the factor loadings appear distinctive enough to mitigate doubts about scale items and factor associations.

Coefficient alpha was also used to test the scales for internal consistency or "reliability." Alpha measures the proportion of a scale's total variance that is attributable to a common source, presumably the true score of a latent variable underlying the items. The reliability coefficients, displayed in Table 2, ranged between 0.63 and 0.91, with an average of 0.82. These results are comparable with Flory et al. (1992). Moreover, DeVellis (1991, 85) describes a coefficient alpha ranging from 0.80 to 0.90 as "very good." Generally speaking, the values for the contractualism and relativism dimensions are lower than the moral equity

Table 2. Reliability Coefficients (Coefficient Alpha)
for Each Dimension and Each Scenario

Dimension	Scenario			
	A	B	C	D
Moral Equity	.80	.91	.87	.87
Relativism	.63	.88	.81	.89
Contractualism	.64	.87	.80	.84

Table 3. Means for the Univariate Ethical/Unethical,
Behavioral Intention, and Multivariate Measures

Scenario	Univariate Means		Multivariate Means[c]		
	Ethical/ Unethical[a]	Behavioral Intention[b]	Moral Equity	Relativism	Contractualism
A	6.56	6.10	6.36	5.25	6.01
B	5.79	4.43	5.52	4.19	5.27
C	6.30	5.19	6.19	5.25	5.85
D	5.18	3.72	4.78	3.75	4.32

Notes: [a] The univariate ethical/unethical measure consisted of a single bipolar scale with "ethical" (scale value = 1) and "unethical" (scale value = 7) endpoints.

[b] Behavioral intention was measured by using a single bipolar scale with the statement, "If you were responsible for making the decision described in the scenario above, what is the probability you would make the same decision?" "Highly probable" (scale value = 1) and "highly improbable" (scale value = 7) were the response options.

[c] All scales were coded from 1 to 7, and summated constructs were averaged to allow for comparison. For moral equity, fair, just, etc. was scaled 1.0 while unfair, unjust, etc. was scaled as 7.0; for relativism, culturally and traditionally acceptable was scaled as 1.0 while culturally and traditionally unacceptable was scaled as 7.0; for contractualism, does not violate an unwritten contract and unspoken promise was scaled as 1.0 while violates an unwritten contract and unspoken promise was scaled as 7.0.

dimension. This is not surprising, given that scales with a smaller number of items produce lower coefficient alphas.

Content Validity

Content validity is an assessment of the construct that is being measured. That is, does the scale actually measure the ethical perception construct? To test content validity, the three dimensions (moral equity, relativism, and contractualism) were compared with an "ethical-unethical" univariate ethics measure.[5] High covariations between the multivariate measures and the univariate measure would suggest that the multivariate measures capture much of what the subjects mean by "ethical." Multiple regression models were estimated for each of the four scenarios using the three factor scores as the independent variables and the univariate measure as the

Table 4. A Comparison of the Multivariate and Univariate Ethics Measures
Univariate Ethics Measure = $ß_0$ + $ß_1$ Moral Equity + $ß_2$ Relativism +
$ß_3$ Contractualism + e

	Regression Results			
Scenario	Overall R^2	Estimated β_1	Estimated β_2	Estimated β_3
A	0.2717	0.4709	0.1311	0.1296
B	0.5671	0.9297	0.4856	0.3724
C	0.5276	0.7749	0.2420	0.2363
D	0.5203	1.0226	0.5279	0.2663

Note: Overall R^2 is the result of a multivariate analysis and is adjusted for degrees of freedom. All results were
statistically significant at $p < 0.02$. Normalized variables were used to estimate betas. The estimated
betas are as follow:
$ß_1$ = estimated beta value for moral equity dimension;
$ß_2$ = estimated beta value for relativism dimension;
$ß_3$ = estimated beta value for contractualism dimension.

dependent variable.[6] Means for the univariate ethics measure and multivariate measures are shown in Table 3 and regression results appear in Table 4.

The three dimensions explain between 27 and 57% of the variance in the
univariate measure. For all scenarios, R^2 was significant at $p < 0.02$. Despite this
statistical significance, the amount of variance left unexplained indicates that
other factors exist in the decision-making process. In all four scenarios the moral
equity dimension exerted the greatest influence, followed by the relativism
dimension, with the contractualism dimension exerting the least influence.

Predictive Validity

Predictive validity is the establishment of an empirical association of an item or
scale with another criterion. Whether or not the theoretical basis for that association is understood is irrelevant for this type of validity. Thus, predictive validity
per se is more a pragmatic issue than a scientific one because it is not concerned
with understanding a process but merely with predicting it. Logically, one may
presume that an individual's ethical response to an observed happening plays an
important role in determining that person's willingness to behave in the same
way. Therefore, a measure of the likelihood that the subject would behave in the
same way as the person depicted in the ethical scenario is the criterion variable
used in this test. We measured the behavioral intent of each subject as did Flory et
al. (1992), obtaining responses to the following statement, "If *you* were responsible for making the decision in the scenario above, what is the probability you
would make the same decision?" The range of possible responses was from one
(Highly Probable) to seven (Highly Improbable). Table 3 presents the means for
the behavioral intention variable, and Table 5 presents the variable's regression

Table 5.   A Comparison of the Multivariate and
Behavioral Intention Ethics Measures
Behavioral Intention Measure
$= \beta_0 + \beta_1$ Moral Equity $+ \beta_2$ Relativism $+ \beta_3$ Contractualism $+ e$

| | Regression Results | | | |
Scenario	Overall R^2	Estimated β_1	Estimated β_2	Estimated β_3
A	0.2505	0.5553	0.4395	0.1560
B	0.4857	1.1301	0.8771	0.4615
C	0.3989	0.7987	0.9110	0.2679
D	0.5338	1.0420	1.2622	0.4427

Note: Overall R^2 is the result of a multivariate analysis and is adjusted for degrees of freedom. All results were statistically significant at $p < 0.04$. Normalized variables were used to estimate betas. The estimated betas are as follow:
β_1 = estimated beta value for moral equity dimension;
β_2 = estimated beta value for relativism dimension;
β_3 = estimated beta value for contractualism dimension.

results. Multiple regression models were estimated for each of the four scenarios using the three factor scores as the independent variables and the behavioral intention measure as the dependent variable.[7]

The scale's three dimensions explain between 25 and 54% of the "behavioral intention" variance. The behavioral intention mean scores indicate a greater likelihood of engaging in activities specified in scenarios B and D than scenarios A and C; subjects were more inclined to participate in those activities that they ranked as less severe forms of unethical behavior. The individual bivariate beta values are statistically significant at the 0.04 level. The least influential dimension of all scenarios was the contractualism dimension, as depicted by smaller beta values. In scenarios A and B, the moral equity dimension is more influential than the relativism dimension. The relativism dimension is more influential than the moral equity dimension in scenarios C and D. These results satisfy the expectations for predictive validity.

A comparison of the multivariate to the univariate measure's ability to "explain" behavioral intent provides additional evidence of the usefulness of the multivariate ethics measure. The behavioral intention variance explained using only the univariate measure is 0.04 for scenario A, 0.23 for scenario B, 0.12 for scenario C, and 0.19 for scenario D. In each case, the amount of variance explained is reduced by more than half when the univariate measure is used in lieu of the multivariate ethics measure, demonstrating that the multidimensional measure increases our understanding of individual intent beyond what we learn from a single-item measure.

It is interesting to note some observations provided by a comparison of Table 4 with Table 5. First, with the exception of scenario D, the multivariate measure explains a marginally higher proportion of variance for the univariate ethics measure than for the univariate behavioral intention measure. It appears that the behavioral intention of an individual is influenced by more than the ethical judgment of a situation. This appears to be a point in the decision making process where a balance is sought between ethical judgment and other competing pressures. For example, Robin, Reidenbach, and Forrest (1996) concluded that the link between ethical judgment and moral intention is influenced by one's perceived importance of the ethical issue (PIE). The link between ethical judgment and behavioral intent is weakened at low levels of PIE (i.e., an issue is perceived to be relatively unimportant) because of the relative increase in the importance of these competing pressures.

Second, the multivariate measure explains less of the behavioral intention variance for scenarios judged by respondents to be more serious ethical breaches (scenarios A and C). It is interesting to note that this is consistent with the findings of Flory, Phillips, Reidenbach, and Robin (1992, 295). Scores on their univariate ethics measure were 5.8, 6.1, 5.8, and 4.8 for four ethical scenarios (ethical = 1.0); i.e., one scenario was judged substantially more ethical (less unethical) than the other three. The proportion of variance accounted for by the multivariate ethics dimensions was twenty points higher for the scenario judged least unethical (univariate ethical/unethical mean of 4.8) than the mean proportion of variance accounted for in the other three scenarios. An implication in these studies, for the scenarios depicting more unethical breaches of conduct, is that something else besides the ethics of the situation played an important role in influencing behavioral intent. Or, generally speaking, perhaps the link between ethical judgment and behavioral intention is weakened for more serious breaches of ethical conduct.

The tests for construct validity, content validity, and predictive validity provide evidence that the instrument developed by Flory et al. (1992) is robust to different subjects and ethical situations. Therefore, the generalizability of the technique was further documented by our study. The test for construct validity confirmed the three anticipated factors (moral equity, relativism, and contractualism). The content validity test demonstrated that the three dimensions appear to measure the ethical judgment construct. The predictive validity tests indicated that the multidimensional scale increased our understanding of individual intention beyond what we can learn from a single-item measure.

DISCUSSION

A popular stream of ethics research, using Kohlberg's (1969, 1984) stage-sequence model of ethical cognition, is discussed in this section and is incorporated with the current study's results to form recommendations to accounting academics who are involved in ethics education. First, it is important to note that research, such as the studies by Ponemon and Glazer (1990) and Green and Weber (1997), described earlier, provided evidence that the ethical development of accounting students can be influenced by their education. Both studies employed Lawrence Kohlberg's (1969, 1984) stage-sequence model of ethical cognition. The model defines a series of three cognitive levels (each consisting of two stages), somewhat akin to the rungs of a ladder. All individuals move upward through these developmental levels beginning at "pre-conventional morality," to the second level termed "conventional morality" and sometimes to the final level called "post-conventional morality." At the pre-conventional level, a person responds to notions of "right" and "wrong" essentially in terms of the punishments or rewards associated with actions. A socialization process takes place at the conventional level in which one identifies oneself in relation to others. That is, maintaining the expectations of one's family and other collections of individuals are perceived as valuable. The post-conventional person differentiates him/herself from the rules and expectations of others and defines his or her values in terms of self-chosen principles.

It is interesting to note that research has consistently shown the stage of moral (ethical) development of professional accountants (as measured by a "P" score) to be lower than the average college graduate.[8] P scores obtained in studies of practicing accountants are shown in Table 6. Also, two studies, Ponemon (1990) and Ponemon and Glazer (1990) reported that the variability of P scores was significantly lower for accounting alumni than for accounting students, which is indicative of a socialization process within the accounting profession and accounting firms. That is, the member selection and advancement process used in the accounting profession yields increasingly homogeneous levels of ethical reasoning (Ponemon 1992). In all, it appears that a socialization process exists that results in professional accountants not reaching the level of moral development attained by average college graduates and other groups of professionals.

These findings may be surprising to readers who are familiar with the high ranking of ethical standards accorded accountants. In various polls and surveys, accountants have been ranked at or near the top by virtually all subgroups of respondents (see, for example, Pinnacle Group 1989; Touche Ross 1988; Harris

Table 6. A Comparison of College Graduate and Practicing Accountant "P" Scores

Study and Date	Sample Size	Average P Score
Practicing Accountants		
Armstrong (1987)	174	38.1
Ponemon (1987)	180	38.1
Ponemon & Glazer (1990)	43	43.6
Lampe & Finn (1994)	207	40.0
Average College Graduates		
Rest (1986)	885	44.8

1986). However, upon closer examination this is not inconsistent with the findings of accounting research employing the stage-sequence model. Accounting
codes of conduct are rule oriented, which is supported and rewarded by public
opinion and market forces. Noreen (1988) conducted an agency theory analysis of
the market for public accounting services and concluded that an economic advantage is gained because of the public's perception of high ethical standards and
strict adherence to rules by the accounting profession. Higher levels of ethical
(moral) development in the accounting profession could lead to less structured
ethical decision making, thus weakening this public perception. Lampe and Finn
(1994) found that students with higher P scores (post-conventional reasoning)
made more liberal decisions—i.e., they knowingly deviated from ethical rules and
norms more often than did students employing more conventional reasoning.
Thus, they questioned whether development to the highest stages of post-conventional reasoning was beneficial (Lampe and Finn 1994, 115).

It appears, therefore, that students would be well served if their accounting education facilitated adoption of the mores of the accounting profession. Mautz
(1975) suggested that educators should attempt to develop professional beliefs
and attitudes among accounting students. Similarly, Mayer-Sommer and Loeb
(1981) argued that school experiences should do more to provide students with a
full sense of professional identity, an appreciation of their ethical and legal duties,
and an understanding of the profession's demands and risks. A student's growth
to the conventional, and perhaps early post-conventional, stages of moral reasoning would serve to aid the student's socialization into the accounting profession.

Growth to the conventional stages of moral reasoning fits with the relativism and
contractualism dimensions tested in this paper. That is, ethical decision making is
influenced by parameters inherent in one's social/cultural system as well as the
rules implied in the system. Hence, a necessary first step in the ethical education
process is the enhancement of students' knowledge of the behavioral norms that
drive the profession's members. Knowledge of the AICPA Code of Professional
Conduct will build a foundation for students, enabling them to identify with the
profession and to take its values as their own. Also, in developing solutions to ethical dilemmas, solutions that are consistent with the ethical standards developed
by the accounting profession, one must first know and understand the standards.

Aiding a student's moral growth beyond the conventional, to the early
post-conventional stage is also desirable. By enhancing moral development to
higher stages of principled reasoning one will be better equipped when confronted with situations which are ill defined by known (e.g., rules explicitly
stated in the AICPA Code of Professional Conduct), or which appear to have
conflicting rules. Development beyond decision making based simply on memorization and application of a code of conduct, to application of social norms such
as fairness and right and wrong, will better equip students when confronted with
ethical dilemmas poorly defined by expressly stated normative guidelines. The
development of this form of reasoning is consistent primarily with the relativism

and contractualism dimensions of the ethical decision making construct. Implied duties and an understanding of right and wrong embedded in one's social and cultural systems are inherent in these dimensions. A greater understanding of these basic values from an accounting perspective will enable students to better recognize moral issues and frame ethical judgments in difficult ethical situations. Advancement to the post-conventional stage may also apply to the moral equity dimension. One's self-chosen values (post-conventional) are a function of one's fundamental beliefs about right and wrong human conduct (moral equity). To the extent that the student's fundamental beliefs of notions such as "good," "bad," and "virtuous" can be influenced in a professionally acceptable way, the moral equity dimension of ethical decision making will be positively influenced.

According to Rest (1986), programs shown to be most effective in raising the level of post-conventional decision making consist of treatments involving participant discussion of controversial moral dilemmas (e.g., case studies). It has been argued that students do not significantly increase moral development by merely learning to obey rules promulgated by others. Higher levels of moral development are best attained when there is cognition about ethical issues, alternative solutions are discussed, and freedom is provided for the exercise of judgment to select the alternative considered best. In advocating a case based approach to ethics education, Lampe and Finn (1994, 120) state that, "Teaching an auditing student that it is wrong to break the rule of confidentiality is far different from discussing reasons for client confidentiality, considering dilemmas involving confidentiality, and providing opportunities for students to independently decide what is right in different confidentiality dilemmas that have varying impact on multiple audit-stakeholders." However, students should be encouraged to give serious consideration to the values of the profession that underlie moral behavior in accounting, as opposed to totally self-chosen values that may be in conflict with the ethics of the profession.

CONCLUSIONS

This study had two stated purposes. One was to test the multidimensional measure developed by Flory et al. (1992) in a new type of ethical setting with a different type of experimental subjects. The objective was to test the generalizability of the measure. We found that the dimensions of moral equity, relativism, and contractualism capture the construct of ethical judgment of accounting students as was demonstrated in earlier studies of professional accountants and advertising managers. Tests also indicated that these three dimensions explain a substantial portion of the variance in the determination of whether or not an action is judged to be ethical.

The other purpose of the study was to draw inferences for pedagogical purposes. We found that in our measure of the ethics construct, moral equity was the most important dimension, followed by relativisim, then contractualism. Growth to the conventional levels of moral reasoning fits with the relativism and contractualism dimensions. The parameters inherent in one's social/cultural system as well as the rules of the system influence ethical decision making. Therefore, the ethical education process should increase students' knowledge of the behavioral norms that drive the profession's members. Increasing a student's moral growth to the formative, post-conventional stage is also desirable. At this level, one will be better prepared to confront situations not defined by rules or which appear to have conflicting rules or norms. The post-conventional stage fits with all three dimensions examined. However, care should be taken to discourage students from using totally self-chosen values in lieu of conventional morality.

Several limitations should be considered. First, data was gathered from seven universities and the subjects were not chosen randomly. Therefore, the results of this study may not be generalizable to other universities. Second, the factor analyses were conducted using a principal component analysis. This statistical technique transforms the raw data into new measurement variables with a mean of 0 and standard deviation of 1, which allows the researcher to compare the relative effect on the dependent variable of each independent variable (Hair, Anderson, Tatham 1987, 20). A common unit of measure is created, and the coefficients tell us which variable is most influential. However, the beta values can be interpreted only in the context of the other variables in the equation. For example, we find that the beta value for moral equity is larger than the beta value for relativism and contractualism. Therefore, value judgments about the three dimensions in the model may be made in a relative sense, not in an absolute sense. Furthermore, a substantial portion of variance is left unexplained by our models indicating that there are other, undiscovered influences of ethical decision making. Last, the amount of variance explained for behavioral intention changes substantially when the severity of the act increases. A plausible explanation is that other influences become important with an increase in severity. Identifying these additional pressures would provide insights into the behavioral intention of individuals in making ethical decisions.

Research opportunities exist in this area. Others could replicate our study by comparing results between private and public institutions. Programs with a strong liberal arts foundation could be compared to those with traditional accounting programs. Our results suggest that the multidimensional measure explains substantially less behavioral intent for ethical breaches considered more serious by experimental subjects. Do other dimensions exist when a scenario is judged more unethical? Why do the dimensions employed in this study become less important when a scenario is judged less ethical? Such an understanding is crucial to our ability to foster ethical behavior in those we teach.

APPENDIX

Scenario A

Hank Toms is a senior accounting major at a major state university, and has an excellent GPA which he very much wants to maintain. Hank has fallen behind in his course work. He is barely maintaining his GPA and certainly cannot complete the term paper assignments in two of his courses. A fellow student told Hank that he can purchase completed papers for the courses from a commercial research firm. Hank knows that submitting such papers are a violation of university policy and could cause his suspension. However, not purchasing the papers will probably result in Hank's GPA decreasing substantially.

Action: Hank decides to purchase the papers from the commercial research firm. Please evaluate this action of Hank Toms.

Scenario B

Joel Williams is a junior accounting major at an excellent regional university. He attends the university and works full time to support his wife and infant daughter. Joel's full time job and the heavy burden of preparing for an accounting career have caused his grades to be marginal. Joel is having particular trouble with the second intermediate accounting class and is barely passing. The university requires that all accounting majors make a C or above in upper-level accounting courses. A friend of Joel's takes the class in the morning and has offered to tell Joel what questions are on the exam before Joel takes the exam in the late afternoon. Joel is convinced that such information would allow him to pull his grade up from a low D to a C, and possibly a B.

Action: Joel decides to accept his friend's offer and find out what questions are on the intermediate accounting examination. Please evaluate this action of Joel Williams.

Scenario C

Barbara Bryant is pursuing a master's degree in accounting at a major state university. She received a graduate assistantship in the accounting department to help pay her bills. She works 20 hours a week and makes copies of tests and performs various other routine duties as needed. Barbara excelled as an undergraduate major but has found graduate school to be much more difficult. She has already made two C's in graduate school and one more grade of C will mean automatic expulsion from the school. She is currently taking a theory course that is very difficult and she barely has a B average.

One day in the copy room Barbara noticed that some smudged copies of the next examination for her theory class had been thrown in the trash. Out of curiosity she briefly looked over the examination and realized she could not pass it. However, if she took the examination home and started studying it she could make an excellent grade. Barbara knows that it is a violation of university policy to obtain a copy of an examination prior to it being given.

Action: Barbara decides to take the examination home so she can prepare for the next examination. Please evaluate this action of Barbara Bryant.

Scenario D

Stephanie Cummings is a senior accounting major at a major state university who is expecting to graduate at the end of the current semester. Her professor returned all examinations on the final day of classes so the students could prepare for the final examination.

Stephanie noticed that the professor had made an arithmetic error on her last regular examination in her favor. The mistake was substantial; it was 30 points in her favor. The 30 points could make a difference in Stephanie's final course grade. The professor has not approved ignoring grading errors in the student's favor but has no stated or written policy on the matter.

Action: Stephanie decides not to inform the professor about the grading error. Please evaluate this action by Stephanie Cummings.

NOTES

1. See, for example Loeb (1988, 1990, 1991), Langenderfer and Rockness (1989), and Hiltebeitel and Jones (1991).

2. The institutions included two doctoral institutions, three with master of accountancy programs, and two with MBA programs. Four of the institutions were urban campuses and three were rural campuses.

3. In the study by Ameen et al. (1996), students rated the severity of each scenario using the following six-point scale: (0) 'not cheating' (1) 'least severe' (2) 'somewhat severe' (3) 'moderately severe' (4) 'quite severe' (5) 'most severe.'

4. This statistical technique transforms the raw data into new measurement variables with a mean of 0 and standard deviation of 1. These standardized coefficients allow the researcher to compare the relative effect on the dependent variable of each independent variable (Hair, Anderson, Tatham 1987, 20).

5. The univariate ethical/unethical measure, first employed by Flory et al. (1992), consisted of a single biplolar scale with "ethical" (scale value = 1) and "unethical" (scale value = 7) endpoints and it appeared for each scenario. The mean response for Scenarios A and C were 6.56 and 6.30, respectively, indicating they were considered very unethical. The mean response for Scenarios B and D were 5.79 and 5.18, respectively, indicating they were considered less unethical than A and C. This finding is consistent with the results of Ameen et al. (1996).

6. The model was re-estimated with a dummy variable for each of the seven schools for each of the four scenarios to test whether the results were caused by differences in the schools. That is, 28 models were run for Table 4. In only two of these models were significant differences found. Therefore it does not appear that differences in the schools are causing the results.

7. The model was re-estimated with a dummy variable for each of the seven schools for each of the four scenarios to test whether the results were caused by differences in the schools. That is, 28 models were run for Table 5. In only two of these models were significant differences found. Therefore it does not appear that differences in the schools are causing the results.

8. These studies used the psychometric instrument referred to as the Defining Issues Test (DIT), developed by James Rest (1979), which provides a scoring of moral development in terms of Kholberg's six stages. The most commonly referenced measure from the DIT is the "P" score, which provides a measure of the percentage of an individual's post-conventional thinking about what is right or wrong when faced with an ethical decision.

REFERENCES

Ameen, E., D. Guffey, and J. McMillan. 1996. Accounting students' perceptions of questionable academic practices and factors affecting their propensity to cheat. *Accounting Education* 5(1): 1-15.

Armstrong, M.B. 1987. Moral development and accounting education. *Journal of Accounting Education* 5(Spring): 27-43.

Armstrong, M.B. 1993. Ethics and professionalism in accounting education: A sample course. *Journal of Accounting Education* 11(Spring): 77-92.

Beauchamp, T. 1982. *Philosophical Ethics*. New York: McGraw-Hill.

Beets, S.D. 1993. Using the role-playing technique in accounting ethics education. *The Accounting Educators' Journal* 5(Fall): 46-65.

DeVellis, R.F. 1991. *Scale Development: Theory and Applications*. Newbury Park, CA: Sage.

Flory, S., T. Phillips, R. Reidenbach, and D. Robin. 1992. A multidimensional analysis of selected ethical issues in accounting. *The Accounting Review* 67(April): 284-302.

Green, S., and J. Weber. 1997. Influencing ethical development: Exposing students to the AICPA code of conduct. *Journal of Business Ethics* 16(8): 777-790.

Hair, J., R. Anderson, and R. Tatham. 1987. *Multivariate Data Analysis with Readings*, 2nd ed. New York: Macmillan.

Harris, L. 1986. How the public sees the CPAs. *Journal of Accountancy* 162(December): 16-34.

Hiltebeitel, K., and S. Jones. 1991. Initial evidence on the impact of integrating ethics into accounting education. *Issues in Accounting Education* 6(Fall): 262-75.

Jeffrey, C. 1993. Ethical development of accounting students, nonaccounting business students, and liberal arts students. *Issues in Accounting Education* 8(Spring): 86-96.

Kohlberg, L. 1969. Stage and sequence: The cognitive-developmental approach to socialization. Pp. 347-480 in *Handbook of Socialization Theory and Research*, edited by D. Goslin. Chicago: Rand McNally.

Kohlberg, L. 1984. *Essays in Moral Development, Volume II: The Psychology of Moral Development*. New York: Harper and Row.

Lagrone, R.M., R.E. Welton, and J.R. Davis. 1996. Are the effects of accounting ethics interventions transitory or persistent? *Journal of Accounting Education* 14(Fall): 259-276.

Lampe, J.C., and D.W. Finn. 1994. Teaching ethics in accounting curricula. *Business and Professional Ethics Journal* 13(1): 89-128.

Langenderfer, H., and J. Rockness. 1989. Integrating ethics into the accounting curriculum: Issues, problems, and solutions. *Issues in Accounting Education* 4(Spring): 58-69.

Loeb, S. 1988. Teaching students accounting ethics: Some crucial issues. *Issues in Accounting Education* 3(Fall): 316-29.

Loeb, S. 1990. Whistleblowing and accounting education. *Issues in Accounting Education* 5(Fall): 281-294.

Loeb, S. 1991. The evaluation of 'outcomes' of accounting ethics education. *Journal of Business Ethics* 10(February): 77-84.

Mayer-Sommer, A., and S. Loeb. 1981. Fostering more successful professional socialization among accounting students. *The Accounting Review* 56(January): 125-136.

Mautz, R. 1975. The case for professional education in accounting. In *Schools of Accountancy: A Look at the Issues*, edited by A. Bizzell and K. Larson. New York: American Institute of Certified Public Accountants.

Mintz, S.M. 1995. Virtue ethics and accounting education. *Issues in Accounting Education* 10(2): 247-267.

National Commission on Fraudulent Financial Reporting (Treadway Commission). 1987. Report of the National Commission on Fraudulent Financial Reporting (October).

Nazario, S. 1990. Schoolteachers say it's wrongheaded to try to teach students what's right. *Wall Street Journal* (April 6): B1.

Noreen, E. 1988. The economics of ethics: A new perspective on agency theory. *Accounting, Organizations and Society* 13(4): 359-369.

Nunnally, J. 1978. *Psychometric Methods*. New York: McGraw-Hill.

Pinnacle Group. 1989. Ethics survey compares various professions. *PR Reporter* (August 14): 2.

Ponemon, L.A. 1987. A cognitive-developmental approach to the analysis of certified public accountants' ethical judgments. Ph.D. dissertation, Union College of Union University.

Ponemon, L.A. 1990. Ethical judgments in accounting: A cognitive-developmental perspective. *Critical Perspectives in Accounting* 1(2): 191-215.

Ponemon, L.A. 1992. Ethical reasoning and selection-socialization in accounting. *Accounting Organizations and Society* 17(3): 239-260.

Ponemon, L.A. 1993. Can ethics be taught? *Journal of Accounting Education* 11(Fall): 185-210.

Ponemon, L.A., and A. Glazer. 1990. Accounting education and ethical development: The influence of liberal learning to students and alumni in accounting practice. *Issues in Accounting Education* 5(Fall): 195-208.

Reidenbach, R., and D. Robin. 1990. Toward the development of a multidimensional scale for improving evaluations of business ethics. *Journal of Business Ethics* 9(August): 639-53.

Rest, J. 1979. *In Judging Moral Issues*. Minneapolis: University of Minnesota Press.

Rest, J. 1986. *Moral Development: Advances in Research and Theory*. New York: Praeger.

Robin, D., R. Reidenbach, and P. Forrest. 1996. The perceived importance of an ethical issue as an influence on the ethical decision-making of ad managers. *Journal of Business Research* 35(January): 17-28.

Smith, L.M. 1993. Teaching ethics: An update. *Management Accounting* 74(March): 18-20.

Touche, R. 1988. *Ethics in American Business: An Opinion of Key Business Leaders on Ethical Standards and Behavior*. New York: Touche Ross & Co.